Windsor and Maidenhead
D0259800

A Space Traveller's Guide to the Solar System

Also by Mark Thompson

A Down to Earth Guide to the Cosmos

For more information on Mark Thompson and his books, see his website at www.markthompsonastronomy.com

A Space Traveller's Guide to the Solar System

Mark Thompson

BANTAM PRESS

LONDON • TORONTO • SYDNEY • AUCKLAND • JOHANNESBURG

TRANSWORLD PUBLISHERS
61–63 Uxbridge Road, London W5 5SA
www.transworldbooks.co.uk

Transworld is part of the Penguin Random House group of companies
whose addresses can be found at global.penguinrandomhouse.com

First published in Great Britain in 2015 by Bantam Press
an imprint of Transworld Publishers

A CIP catalogue record for this book
is available from the British Library.

ISBN 9780593073339

Typeset in 11.25/16pt Berkeley Book by Falcon Oast Graphic Art Ltd.
Printed and bound in Great Britain by Clays Ltd, Bungay, Suffolk.

1 3 5 7 9 10 8 6 4 2

To my mum and dad, who tirelessly supported
my childhood dream to reach out among the stars.

Contents

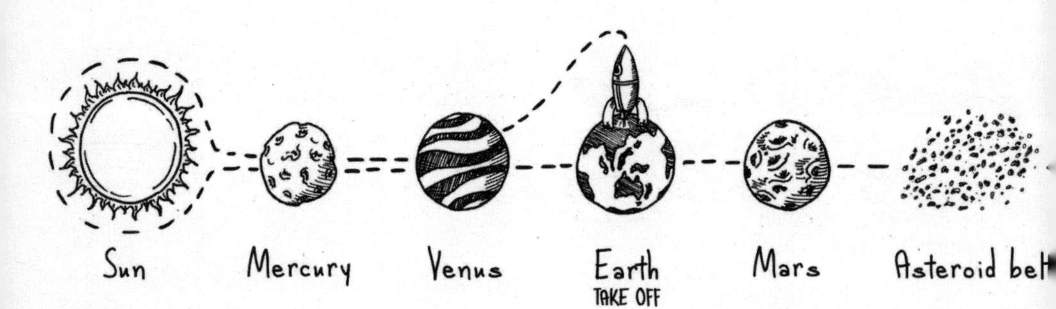
Sun
Mercury
Venus
Earth
TAKE OFF
Mars
Asteroid belt

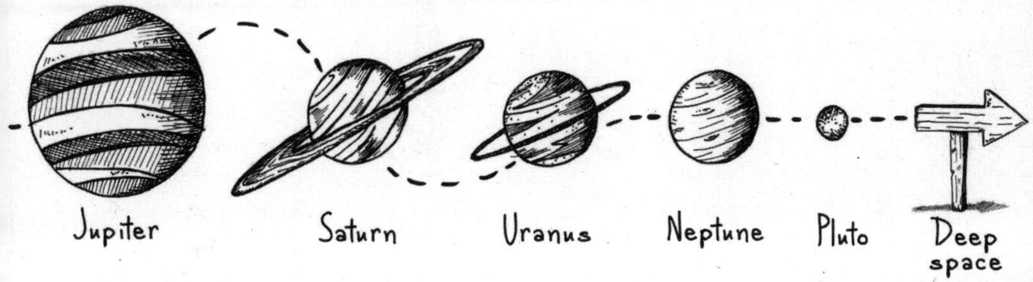

Jupiter

Saturn

Uranus

Neptune

Pluto

Deep space

Introduction

When I was a boy of ten I was taken along to the local astronomy society by my father and on that cold frosty January night almost thirty years ago I saw something that changed my life. As I took my very first peek through a telescope I was greeted by a view of the mighty Saturn, hanging in the inky black depths of space. I could see the delicate hues of the belts on the planet, the rings encircling it, and a moon or two as well. It was an incredible sight, and to this day I have never seen another view that has left quite such an impression on me. During that short encounter it felt as though I had been transformed into an astronaut and was in orbit around an alien world. It was magical. My childhood following that experience was often filled with boyish dreams of becoming a spaceman, and I spent much of my youth working on projects that I hoped would help to fulfil my ambition of flying around the Solar System, taking in strange new worlds and distant moons. But I was not alone. As I was making cardboard rockets, tinfoil space suits and papier-mâché helmets, millions of other children were at home doing the exact same thing.

Dreams of becoming an astronaut have been common among children ever since 20/21 July 1969, when families the world over were huddled around their television sets watching Neil Armstrong and Edwin 'Buzz' Aldrin land on the Moon and become the first men to walk on its surface. Millions were transfixed by the hazy, flickering footage being beamed live from over a quarter of a million kilometres away. Watching those brave explorers, the first humans to set foot on extraterrestrial soil, excited children; they wanted to be those men, to journey into the unknown and explore different worlds. Suddenly space travel was not just the stuff of science fiction but a reality, something tangible and achievable. Even now, decades later, the story of Armstrong, Aldrin and Mike Collins, who stayed in the command module of Apollo 11, continues to inspire countless children who dream of reaching out to the stars.

And these dreams of space exploration aren't just confined to youngsters. As we grow up we may loosen our grip on the desire to become astronauts, but there are plenty of other ways through which we channel our deep-rooted fascination with the alien and the unknown. There are examples scattered through history; poets, authors, artists and musicians have produced many works based on the Universe around us. You only have to think of the beautiful orchestral pieces by Gustav Holst inspired by the planets or the plethora of science fiction stories and films to see that astronomy reaches out to people in a way no other science does. We have always been fascinated by the unfamiliar, largely because it allows our imagination free rein to conjure up all sorts of weird and wonderful fantasies. Alien worlds and their potential inhabitants are the epitome of this, and the dream of space exploration has been a part of human culture for many years. Be it through fear or fascination, it has also proved to be a unique way of bringing people together, as the photographs from NASA's Apollo era demonstrated. Not only did the images give us a new view of the Moon, they also gave us a fresh perspective on Earth. Who can fail to be humbled after seeing pictures of our fragile planet hanging in the dark, cold, lonely depths of space?

Out of my childhood dreams and games grew an interest in astronomy and the Universe that has followed me into adulthood, along with questions and ideas that have been bubbling around in my head for many years. I have always wanted to see my home, planet Earth, from space and to learn more about its relentless orbital dance with the Moon. I dream of heading to Mercury and Venus, and to the Sun to learn how it works and see just how close it is possible to get to those incredible temperatures. I want to know if it is possible to survive one of the sandstorms on Mars, or to navigate through the dense rocks of the asteroid belt as I approach the realm of the gas giants. Would it be possible for me to fly through the eye of Jupiter's Earth-smashing hurricane, which I have seen through a telescope on many occasions, and are the rings of Saturn really made up of billions of pieces of rock and dust?

Approaching these questions from a theoretical perspective, I often give lectures to others like me who are in awe of the Universe. Children are regularly in the audience and it never ceases to amaze me how wonderfully insightful their questions can be. I recall a child of about ten years once asking me, 'What shape is the Universe?' That's a fantastic question to come from an adult, and I was surprised to hear a child ask it. It just shows that we all have probing thoughts and unanswered queries about worlds beyond our own. In some ways, I think the wonderment of children is analogous to that of our ancestors as they slowly became aware of the Universe around them. Sadly, I think a lot of adults tend to lose this fascination with the natural world; but that one question made me realize that trying to understand the secrets of space brings us all together. Certainly everyone else in the room with that ten-year-old boy was waiting for my answer. Whatever your age, we are all on this journey of learning and discovery together.

You may be surprised to know just how much we have come to learn about the Universe from remote observation. Indeed it is fair to say that we have learned more by study through a telescope than we have by direct exploration. But there are limitations, and these are largely a result of our technological capabilities. As technology develops and we discover new ways of exploring the Universe, both by remote observation and direct exploration, our view of the Solar System and the planets will change and evolve. But before you climb aboard your spacecraft it is worth glancing back briefly at the past, to take in some of the important discoveries that have led to our current view of the Solar System.

Wand'rin' Stars?

By 'Solar System', I mean the region of space that is dominated by the influence of the Sun. It includes the planets Mercury, Venus, Earth, Mars, Jupiter, Saturn, Uranus and Neptune and a whole host of comets, asteroids and minor planets. Our distant ancestors knew that these bodies were different to the stars around them because they could see them moving independently. Indeed the term 'planets' comes from the Greek word *planes*, which literally means 'wanderer', and that nicely describes how they seem to move aimlessly against the background of the stars. For many centuries it was believed that the Earth was at the centre of this system, with the Sun, Moon, planets and stars all in orbit around us. This model was known as the Geocentric Model and it was articulated in a piece of work by the Greco-Egyptian scholar Claudius Ptolemaeus, or Ptolemy. However, more detailed observations of the motions of the planets, for example the strange backward or retrograde loops of Mars, revealed that the Earth could not be at the centre of the Solar System. An alternative model was suggested as early as 3 BC but the idea received no real support until the sixteenth century when Nicolaus Copernicus suggested the notion of a Sun-centred system.

I Can See (Relatively) Clearly Now

The invention of the telescope in 1608 by the Dutch spectacle maker Hans Lippershey was the real turning point in our ability to study objects in the Solar System, and therefore understand their origin. With this wonderful new invention and its slightly modified successors, astronomers uncovered some of the great mysteries of our planetary cousins, but the discoveries were always limited. The attempt to see any fine detail was restricted by the conditions in our own atmosphere through which the incoming light had to pass. Not only was cloud a problem but the amount of dust and pollution in the atmosphere grew as industry slowly developed. All this as well as a very unstable atmosphere often made the image jump around, as though being viewed through a pan of boiling water. The only solution was to get up close, and to do that required getting telescopes, and indeed people, above the Earth's atmosphere.

Rocket Man

The technology needed to travel into space started to emerge long before any thoughts of actually doing so. Many people will be familiar with the experimentation of the Chinese with gunpowder, but the first signs of rocket propulsion appeared around 400 BC when a Greek by the name of Archytas suspended a water-filled wooden bird on a string above a fire, the escaping steam providing the propulsion. In its simplest form, this is the first documented appearance of Newton's law of motion, which dictates that every action has an equal and opposite reaction. A few hundred years later, another Greek known as Hero who lived in Alexandria produced what could be considered a precursor to a motor, based on the concept of action and reaction. A fire was lit under a bowl of water which had two vertical tubes above it. The tubes collected and channelled the steam into a sphere which had two L-shaped tubes on opposite sides, allowing the steam to escape and causing the sphere to rotate.

It is difficult to pin down the original appearance of a chemical rocket but the Chinese were certainly among the earliest to experiment. It is thought that they may have realized the potential of such rockets during religious festivals when throwing bamboo sticks filled with a gunpowder-like substance on to fires to create explosions. There is a good chance that rather than explode, some of them will have shot off out of the fire, giving those who watched the idea of using chemicals to propel objects. They started experimenting with arrows attached to bamboo shoots filled with gunpowder in order to propel the arrows further when fired from a bow. It is believed that they then realized they could dispense with the bow and just fire the arrows using rocket power alone.

The first use of what might be considered the precursor to the modern rocket was during the battle of Kai-Keng in 1232, when the Mongols were at war with the Chinese. The Mongol army was held back by the Chinese who used what was later recorded in history books as 'arrows of flying fire'. What seemed frightening to the Mongols was a simple form of rocket – a tube blocked off at one end and filled with gunpowder. Once the gunpowder was ignited, the 'flames' escaped out of the open back producing thrust which propelled the rocket forward. Directional control was maintained by a long stick that protruded out of the back of the rocket to keep it going forward.

There were many subsequent developments and enhancements of these primitive rockets but perhaps the most relevant for space exploration was made by the German firework manufacturer Johann Schmidlap some time in the 1500s. He invented what became known as the 'step rocket', which was the forerunner of rockets like Saturn V, which launched Armstrong, Aldrin and Collins to the Moon in Apollo II in 1969. Schmidlap's invention was a multi-staged firework which used a large rocket to carry a second smaller rocket higher in the sky so that when the first rocket engine came to the end of its burn, it ignited the second rocket. Of course at the time this was used only for firework displays but the concept was later used, and still is today, for launch vehicles.

A wonderful sixteenth-century Chinese legend suggests that the first human to attempt transportation by rocket was one Wan Hu. He was said to have constructed a rocket-powered chair that was suspended under a kite that had around fifty Chinese 'fire rockets' built on to it. At his command, Wan Hu's assistants lit the rockets and, so the legend goes, there was a tremendous roar and clouds of smoke, but by the time the smoke cleared, Wan Hu had vanished. His body was never found and no one really knows what happened to him, although it is likely that instead of racing off into the heavens on his rocket-powered chair, the contraption exploded, blowing Wan Hu to pieces. Thankfully modern rocket flight is a little safer than that, although there have of course been tragedies. In the first fifty years of space flight eighteen astronauts have died in four separate accidents: in 1967, a parachute failed to open on Soyuz 1 after it re-entered the Earth's atmosphere; Soyuz 11 decompressed in 1971 just after undocking from the space station; in 1986 the *Challenger* space shuttle disintegrated after launch; and in 2003 the *Columbia* shuttle disintegrated on re-entry.

Space exploration has come on in leaps and bounds since the rocket was first invented. In 1961 cosmonaut Yuri Gagarin became the first person to complete an orbit of Earth, and just eight years later the historic Moon landing occurred. Then focus turned to the development of some kind of habitable space station, which differs from a spacecraft in having no form of major propulsion or landing system. The first space station was Salyut-1, which was launched in 1971 and, like its early successors, was designed to be sent up in one piece. Things changed in 1986 with the launch of Mir and then the International Space Station (ISS), both of which were modular systems, designed to be launched in stages and built in space. At the turn of the twenty-first century the ISS was the only operable space station in orbit, which has now been permanently manned for over fourteen years.

Don't Stop Believin'

Those individuals who have travelled into space have by and large been highly trained and carefully selected. But that doesn't stop the rest of us from dreaming. From Buzz Aldrin to Buzz Lightyear we venerate astronauts as heroes, as people we respect and often try to imitate. So how exactly can the rest of us follow in their footsteps? Clearly, before space travel can become routine things need to change: it needs to become easier and safer and it also needs to become cheaper. But we are well on the way. With the recent birth of space tourism a seat on a sub-orbital space flight costs in the region of £150,000. The chance to go into space properly – and by properly I mean experience orbits of Earth and hours if not days of weightlessness rather than just take a trip up and straight back down again – costs in the region of £20 million. That's a lot of pocket money. Both of these may be unobtainable for the average person in the street, but these commercial space programmes none the less mark the first time untrained members of the public can buy their way into space. It is a giant leap in a very interesting direction, and who knows where it will take us in years to come.

For now, though, most of us will have to seek solace in the power of the imagination. I still think back to my childhood and my fantasy of becoming a 'spaceman', coloured as that vision now is by real-life missions, reports and discoveries. Even with all the things I have now learned about space and the Universe, my mind still wanders as I picture myself flying around the Solar System, visiting each of the planets in turn. So what would such a journey be like? What would one see and feel and experience? This book is about that journey.

We'll start by looking briefly at the mechanics of the Solar System and our knowledge of planetary movement, which will allow us to plan a route and prepare for the journey to come. Leaving the Earth and Moon behind, your voyage will first take you to the inner Solar System to visit the Sun, then Mercury and Venus, before swinging past the Earth again on your way out of the inner Solar System. After passing the red planet Mars you'll head through the asteroid belt and beyond. The first gas giant planet you'll encounter is the largest in the Solar System, Jupiter. Saturn, Uranus and Neptune are next on the itinerary before you immerse yourselves in the mysterious objects that live in the depths of interplanetary space.

This book is an extension of my dream, and the dreams of many others, to see if at least in theory it is possible to fly around the Solar System and visit alien worlds. Join me now and prepare to depart on your very own space mission.

ONE

Flight Planning

As with any journey, before you can set off on your space odyssey you must first decide exactly where you're going and how you're going to get there. You need a flight plan. Journeys on Earth are pretty easy to plan and execute because we have mapped the surface quite extensively. Generally speaking, nothing moves and everything stays in the same relative position, which makes navigation pretty simple. In a car, for example, one needs only to follow the correct roads and the towns and cities we wish to reach will be at the end of them. Even in an aeroplane, once you've accounted for the Earth's curvature, it is relatively straightforward to point in the direction you wish to go and you will get there. It is not quite that easy when it comes to travelling in space.

The real stumbling block when travelling around the Solar System is that it takes a significant period of time to get anywhere, so the relative position of an object will change. If you want to go to Jupiter, for example, if you simply aim your spacecraft at it, by the time you arrive it will have moved somewhere else. What you actually need to do is figure out where the planet will be at the calculated arrival time and aim for that spot instead. It is much like clay pigeon shooting: you have to aim at the place where the clay pigeon will be when the shot arrives, otherwise you miss. You must essentially intercept the moving object. Things get even more complicated for an epic journey like yours, as visiting more than one planet requires complex calculations and, ultimately, complex trajectories. But we can look back at the Voyager and Pioneer missions and learn from their calculations to help you plan your own flight path.

Over the years we have gained a very good understanding of movement within our Solar System which means at any given time we know where all the major bodies, and a great many of the minor bodies, are. Surprisingly perhaps, most of our knowledge about how the planets move comes from naked-eye observations over many hundreds of years. As we saw in the introduction, when mankind first started looking at the sky it became clear that a handful of 'stars' were unlike the rest, and these became known as the planets. Careful study helped to unlock the secrets of how they moved around the Solar System, revealing that their motion was not fully described by the Geocentric Model outlined by Ptolemy. The model was modified, and instead of the planets moving around the Earth in circular orbits, the idea of epicycles and deferents was proposed. In this system, all the planets orbited around a smaller circle known as an epicycle which itself completed a larger orbit around the Earth called the deferent. This went some way to explaining the strange wanderings of the planets, but there were still some anomalies.

In 1543, Nicolaus Copernicus proposed a different idea, which found no favour with the Catholic Church. In his new helio-centric model the Sun had found its way to the centre of the Solar System replacing the Earth, which became just one of the family of planets in orbit around it. It was a move that put Copernicus in direct conflict with the Church which believed at the time that the Earth was the most important object in the known Universe. They believed that it had been made by God and therefore *must* be at the centre of everything, in the most important position possible. Observations didn't support that concept, but the idea of an Earth-centred Universe did not start to lose credibility until 1609, when Galileo published the initial observations he'd made with the newly invented device known as the 'telescope'. In his work *Sidereus Nuncius* he reported craters on the Moon, the rings around Saturn and the four moons of Jupiter, and it was these observations that finally pushed the Catholic Church into a corner. They responded by accusing Galileo of heresy and he was placed under house arrest until he died. It wasn't until 1992 that the Church finally released an official apology for this, several hundred years after his death.

In the same year that Galileo made his first observations, another astronomer, Johannes Kepler, had been analysing observations of the positions of the planets made by his now deceased tutor, a man named Tycho Brahe. This work led to Kepler publishing the first two of his well-known laws of planetary motion.

Kepler's first law states that all planets move in elliptical orbits with the Sun at one of the points of focus of the ellipse. An ellipse is essentially a squashed circle, and you can imagine the two points of focus of the ellipse if you visualize a point at the centre of a circle. Now think about squashing the circle from the top and bottom and the central dot splitting in two and being moved outwards in different directions. In the case of the planets in the Solar System, the Sun is found at one of these points, and it is this that they all appear to orbit. His second law states that a line joining the Sun to a planet (known as the 'radius vector') sweeps out across equal areas of space over equal intervals of time. In other words, planets move faster when they are nearer to the Sun and slower when further away.

The last of Kepler's laws, published ten years later in 1619, explains that there is a mathematical relationship between the time it takes for a planet to complete an orbit and its distance from the Sun. In Kepler's words, 'The square of the orbital period of a planet is directly proportional to the cube of its mean distance from the Sun.' This is a really useful relationship because we can measure how long an object takes to orbit the Sun simply from observation, and knowing that, we can calculate its average distance from the Sun with some accuracy. The same law applies to any other satellite objects, for example the moons of Jupiter. If we measure the time it takes for one of the moons to go around the planet, we can calculate its distance from the planet itself.

Kepler's three laws helped to build a clear picture of the Solar System and allowed us to predict the movement of the planets, and this will be vital information for your journey. But Kepler is not the only figure in history who has helped us understand our planetary neighbourhood.

On 4 January 1643, Isaac Newton was born in a tiny village called Woolsthorpe-by-Colsterworth in Lincolnshire, England. Newton became one of the foremost scientific minds of all time with a real flair for physics and mathematics. The pinnacle of his career came in 1687 with the publication of *Philosophiae Naturalis Principia Mathematica* which included his law of universal gravitation. The law simply states that there is a force between any two objects in the Universe, and that force is defined by the mass of the two objects and the distance between them. Put more accurately, the force is calculated by multiplying the masses of the two bodies and then dividing the answer by the square of the distance between them. This means that if two objects retain the same distance between them but increase in mass, then the gravitational force between them will increase. Similarly, the force will increase if the masses are kept the same but the distance between them decreases. Not only does this describe the forces that caused the apple to fall from a tree in Newton's garden which supposedly led him to 'discover' gravity, it also explains how the Moon is kept in orbit around the Earth, the Earth in orbit around the Sun, and even why the stars orbit the centre of the Galaxy. In fact, gravity (or 'gravitas' as Newton called it, which came from the Latin meaning 'weight') pervades the entire Universe, even binding galaxies into great clusters.

In his works, often shortened to *Principia*, Newton also articulated his three laws of motion. The first is the law of inertia which relates to an object's resistance to changing its state of motion. The law states that any object in motion will stay in its current state of motion without accelerating until acted upon by an outside force. This is of great importance when it comes to space exploration as a spacecraft in flight will continue in its current direction and speed with zero fuel use unless it is acted upon by an external force, which might be the gravity of a planet, a meteoroid strike or its own rocket engines.

The second law explains that the speed at which an object

accelerates will depend on its mass and the force exerted upon it. More accurately, the force referred to is the net force, which accounts for the fact that if you apply a force of 10 units to propel a spacecraft in one direction but simultaneously apply a force of 5 units in the opposite direction the craft will only accelerate by a value of 5 units. This is the net force and is the force Newton considered in his second law. The law gets a little more complicated when you consider the mass of the object in question, as increasing or decreasing the object's mass will have an inverse effect on acceleration!

Newton's third and final law is probably the most widely known: for every action there is an equal and opposite reaction. This means that for every interaction between two objects, there is a pair of matching but opposite forces acting upon the two bodies. There are some wonderful examples of this in nature, such as a bird in flight. Birds fly by flapping their wings, but for every beat of their wings downward, they exert a force on the air towards the ground. The interaction is between the bird and the air so there must be equal but opposite forces. With the bird pushing down on the air, there is an opposite force from the air, pushing the bird upward. The forces in play are equal, but the direction is opposite, and it is this that keeps the bird in the air.

With the amazing and insightful work of Kepler, Newton and a few other scientists along the way, the foundations were set for mankind's exploitation of rocket flight and exploration of the Solar System. Your journey has been made possible by many of their discoveries.

In order to be able to travel to all the planets in the Solar System it is necessary to make use of their gravity to affect changes in the velocity (speed and direction) of your spacecraft. Getting to the inner planets is actually pretty easy because a spacecraft that leaves

Earth's orbit and heads towards the Sun will be accelerated by the Sun's gravity. This can make a fly-by of the inner planets really quite simple as long as the trajectory is right. Heading to the outer planets is a little more tricky, though, because any flight away from the Sun will be decelerated by its gravitational pull, causing your spacecraft to slow. Without any form of assistance it would take an incredible amount of fuel to get to the outer planets. So much, in fact, that it would be impossible simply to launch from Earth. This is because even for small changes in the trajectory of a spacecraft travelling at high speeds a large amount of fuel is needed. The more fuel on board, the heavier that spacecraft becomes, and the heavier it is, the more fuel you need to lift off in the first place – and so continues the problem. This is where the gravity of planets can be used to work in your favour, to assist you on your journey.

Gravity assist, or a 'gravitational slingshot' as the manoeuvre is known, was first used by the Mariner 10 mission that was launched to Mercury and Venus back in 1973. It has been used successfully on nearly every interplanetary mission since, including the historic Voyager and Pioneer projects. Another great example is the Cassini mission, which travelled to Saturn after acquiring the necessary speed by flying first around Venus (twice), Earth and Jupiter before arriving at Saturn. Similarly, the Mercury-bound Messenger spacecraft used the gravity of Earth, Venus and Mercury (three times) to reduce its velocity before it could drop into orbit at the innermost planet. The principle is simple. If you fly a spacecraft through the gravitational field of a planet then the spacecraft and planet will exchange energy. Depending on the mechanics of the fly-past, the spacecraft will either gain energy and speed up while the planet will lose a tiny amount, or vice versa. In the case of your grand journey around the Solar System, the purpose of a gravitational assist will be not only to adjust the direction of the spacecraft but also to speed it up.

Now there is a complication with this concept and it is articulated in the universal rule of the conservation of energy, which tells us that the total energy of a given system must always remain constant. So a spacecraft flying through the gravitational field of a planet should speed up on the approach but then slow down as it moves away again. For the conservation of energy to be upheld, this must be the case, and if observed from the point of view of the planet, this is indeed the case. Despite this apparent contradiction, planetary fly-bys are still of use to space missions. The planet is stationary when we consider the encounter from the planet's point of view. However, the planet has significantly more mass than the approaching spacecraft so, on the approach, the spacecraft speeds up as it gains energy but it then loses the same amount of energy and slows down as it moves away. What does change is its direction, so the closer it flies to the planet, the greater the change in direction of the spacecraft, with the greatest possible direction change being 180 degrees as the spacecraft heads back in the direction it just came from. The total energy in the system is conserved but still, we have only changed direction, not speed.

Let us now consider the encounter from the point of view of the Sun. As we watch, the planet is moving, not stationary as it was when we considered it from the planet's point of view. As the spacecraft swings past the planet, just like before, it speeds up when viewed from the Sun but it steals that energy from the orbital speed of the planet. This is how the speed increases from the point of view of the Sun, mathematically by adding the velocity of the planet to the velocity of the planet *and* the velocity of the spacecraft.

This may be a difficult concept to grasp, but a good way to

imagine it is to think of a tennis player hitting a ball back to her opponent. If you were playing a Grand Slam champion then your shot might approach the champion at 20 kilometres per hour. She is much better than you at the game though and might swing her racket at a mighty 50 kilometres per hour. The champion's racket will experience the ball approaching at a whopping 70 kilometres per hour, which comes from adding the speed of her racket to the rather feeble speed of your shot. As your opponent strikes the ball and it starts to head back towards you, her racket will still see the ball receding at the same speed that it approached, 70 kilometres per hour, but at the receiving end you will experience a ball travelling at an incredible 120 kilometres per hour. As far as your opponent's tennis racket is concerned the ball approached and then receded at the same speed at the exact moment of impact, when the racket can be considered to be stationary. From your point of view, which can be aligned with the point of view of the Sun during a gravitational slingshot manoeuvre, the champion's racket is moving and after impact the ball speeds up, but the racket will slow down by a tiny amount as a result of the interaction. It is the same for rockets flying around planets and it's why the careful use of gravitational slingshots is such an integral part of space flight around the Solar System.

In order to complete your interplanetary voyage it is necessary to identify a route around the planets that will exploit as many gravitational slingshots as possible, to adjust your trajectory from one planet to the next. Maximizing the number of the slingshot manoeuvres will minimize the amount of fuel needed, ultimately reducing launch weight. There will only be a small number of opportunities for such planetary alignments. It's a concept that was used very successfully during the Voyager missions. Both Voyager

1 and Voyager 2 executed slingshots around Jupiter and Saturn but the timing and mechanics of the Saturnian fly-past were slightly different for each. Following the encounter with Saturn, Voyager 1 was ejected from the Solar System while Voyager 2 went on to study Uranus and Neptune. For your mission to work, it is essential to identify a time when the planets will align perfectly to allow you to visit each of them in turn.

Once we are sure of the correct alignment and have a launch date, we can work out your flight path. A Titan Centaur rocket will set you on your journey, sending you first past the Moon and then on towards the inner Solar System for an encounter with the Sun. You could of course take in the inner planets en route but they will serve as useful fly-bys later on to adjust trajectory and increase velocity to get to the outer planets. Unfortunately you will gain nothing from a fly-by of the Sun in terms of velocity because all trajectories are relative to it. After just six months you will arrive at the closest point to the Sun and within the four months that follow you will fly by Mercury and then Venus. After the first Venus fly-by you will take in another orbit of the Sun before another fly-by of Venus just ten months later that will serve to increase your velocity relative to the Sun by 21,000 kilometres per hour. Following the second Venus encounter you will enjoy a rather emotional final look at Earth with a fly-by that will serve to set up your trajectory to the outer planets. You will arrive at Mars just five months later, and after a further four months you will take on the perilous crossing of the asteroid belt. Beyond the asteroid belt things will really start to slow down as you cruise to Jupiter, Saturn, Uranus, Neptune and finally Pluto, before setting off to explore the depths of the Solar System.

Are there any other giant planets orbiting beyond Pluto and the icy Kuiper Belt at the outer limits of our Solar System? If one does exist then the chances of spotting it are pretty slim, but your journey will continue through the so-called 'termination shock', where the influence of interstellar space starts to challenge the dominance of the Sun. A long way beyond this point your trip will eventually become interstellar as you reach the heliopause and the edge of the Solar System. After years of travelling your exploration around the familiar Solar System will come to an end, but the voyage won't be over yet. As you return home the spaceship will leave the heliopause behind and continue into deep space for a rendezvous with the theorized Oort Cloud, which will take a staggering 1,500 years. The final leg of the journey will take the now unmanned ship to its ultimate destination, Gliese 581, a star in the constellation of Libra at a distance of 20.2 light years, taking some 239,000 years.

Once you have a launch date and a flight plan we need to consider exactly how you're going to make that journey. Rockets are obviously the right tool for the job, but what type of propulsion should we choose, and which is the most suitable spacecraft?

The first rocket engine employed a solid rocket fuel just like that used in the solid rocket boosters of the space shuttles. The ignition of the prepared solid fuel mixture produces the thrust to propel the shuttle upwards. The mixture in solid rocket boosters starts out as a thick liquid that can then be cast into various shapes as it cures. A typical solid rocket will be cylindrical in shape with a hollow tube running almost the entire length of the rocket. The ignition takes place within the hollow tube, and as the fuel burns it spreads outwards towards the casing of the rocket. Interestingly, if the shape of the channel inside the cylinder is changed to increase the surface area then the thrust can be increased. The shuttle boosters therefore have a star-shaped channel running through them to give maximum possible surface area. Solid rocket systems are cheaper to produce than the alternative liquid rocket propellant, but they are not as controllable because once the engine has been ignited it cannot be stopped or restarted.

'Specific impulse' is the term used to articulate how efficient a rocket propulsion system is and describes the force produced from a given amount of propellant over a given time period. Although they are cheap and still popular among military agencies, the solid fuel rocket systems have a low specific impulse. A higher specific impulse means a lower rate of propellant flow is required to produce a given amount of thrust. This is a very important factor as it determines the amount of fuel needed for your mission.

The alternative to solid rocket fuel was first tested by Robert Goddard in 1926 when he invented the liquid-propelled rocket. Instead of the slow and uninterruptible burn of a solid rocket, the new liquid rocket used gasoline and liquid oxygen to produce the required exothermic reaction. This time there was a significant difference though: the two chemicals were stored separately and injected into the combustion chamber. The rate at which the chemicals were injected dictated the amount of thrust produced, so for the first time rocket engines became controllable. The same principle was used on a grand scale in the mighty Saturn V rocket that took Neil Armstrong, Buzz Aldrin and Mike Collins to the Moon. The whole rocket assembly stood about 110 metres tall and weighed in at 6.5 million pounds – and of that, 5.6 million pounds was fuel. At launch the fuel economy was just 17.7 centimetres to the gallon, although that did improve drastically as the mission progressed. Compare that to the fuel economy of an average car – in the region of 44 kilometres to the gallon – and you'll realize how expensive rockets are to run. Saturn V was actually composed of three different stages, all of them needed to get the astronauts up into Earth orbit, one firing after the previous had expired and been separated. Detaching each stage after it is spent is a more efficient way of getting into orbit, otherwise you have to carry the extra weight with you and that in turn means you need more fuel. The comparatively tiny command and service modules that actually went to the Moon sat on top of the Saturn V assembly and used small nozzles with compressed gas to make course corrections on the way. A final liquid rocket allowed them to leave lunar orbit and return to Earth.

The liquid fuel system employed by the Saturn V rockets used liquid hydrogen and liquid oxygen as the propellant mixture which has a higher specific impulse than the solid rocket fuels. Both technologies can generate huge amounts of thrust but have a relatively low specific impulse when compared to a new alternative concept known as the Variable Specific Impulse Magnetoplasma Rocket, or VASIMR for short. The VASIMR system has a much higher specific impulse making it highly efficient over long journeys but it generates very low levels of thrust. Imagine holding a sheet of A4 paper on the palm of your hand. The force the weight of the paper exerts on your hand is the same as the thrust generated by the VASIMR engine.

The principle behind the VASIMR engine is simple, and like all other rocket propulsion systems it exploits Newton's third law of motion. The idea was developed by former astronaut Franklin Chang-Diaz who realized it would be possible to use magnetic fields to direct and channel superheated plasma out of the back of the rocket. This is an approach unlike conventional rockets which ignite chemicals in an exothermic reaction, as we have just seen.

To understand how it works we first need to understand the inner workings of an atom. Atoms are made up of a combination of particles in their nucleus called neutrons and protons. The protons carry a positive electrical charge and the neutrons, as their name suggests, are neutral and have no electrical charge at all. Surrounding the nucleus is a shell of electrons which have negative electrical charge and it is a combination of these and the neutrons and protons in the nucleus that determine the properties of the atom. A hydrogen atom, for example, has one positively charged proton in its nucleus and one negatively charged electron in orbit around it, while a helium atom has two protons, a number of neutrons depending on the type of helium, and two electrons. As you can see, generally the number of protons balances the number of electrons so the net electrical charge is neutral.

If you were to remove an electron from either of the atoms, then it would become positively charged; adding an electron would make it negatively charged. This process is known as ionization, and this is a key concept in the functioning of the VASIMR engine. Inside the rocket, hydrogen that carries no charge is injected into a magnetic field which strips away the electron, ionizing the hydrogen and making it positively charged. The ionized hydrogen is then moved to a second magnetic field where radio waves much like those in your microwave oven are used to heat it to temperatures in excess of 50,000 degrees Celsius. This heating turns the gas into a plasma, which is often referred to as the fourth state of matter, the others being solid, liquid and gas. Plasma is distinctly different from the other three states due to the presence of a significant quantity of charged particles. The plasma is then channelled to a third and final magnetic field which acts like a nozzle to expel the plasma which generates thrust to propel the rocket forward.

One of the great benefits of this type of engine is that although its specific impulse is high, it can be adjusted even in flight. When higher levels of thrust are needed, its specific impulse could be reduced; when less thrust is needed and efficiency is more important – for example during the cruise – it could be increased. These properties make the VASIMR engine an excellent choice for your cruise around the Solar System because they can provide much longer periods of low acceleration than a conventional liquid- or solid-fuelled rocket. And there is one really quite wonderful added advantage to the VASIMR rocket: its use of hydrogen, which is a strategically good choice because it is one of the most common elements in the Universe, so even on your trip around the Solar System extra fuel will be easy to come by should the tanks need to be topped up.

On your trip you will actually make use of all of these rocket systems. To get off the surface of the Earth and into orbit you will utilize the high thrust of the solid- or liquid-fuelled system, then once under way you will look to the new technology of the VASIMR system to help speed your journey. Perhaps one of the most important benefits of such an engine is the low amount of fuel needed, which means you can keep your spacecraft light yet retain the means of providing a highly efficient form of thrust when you hit interstellar space. It is not unusual to utilize more than one different form of propulsion, as demonstrated by NASA's Dawn mission to study Vesta and Ceres, the two largest bodies in the asteroid belt. Dawn was launched in 2007 by a Delta 7925-H rocket which used a combination of both liquid and solid fuel. Once in Earth orbit the Dawn probe was powered by an ion engine, which is very similar to the VASIMR concept.

With the path around the Solar System and the rocket technology identified, there are a few physiological issues to consider before you set off, and this is where choice of spacecraft and equipment becomes important. Living on the surface of a planet with an atmosphere and a magnetic field means we are protected from the harsh conditions in space. As soon as we travel beyond the protective confines of our ecosystem we can expect no air to breathe, no atmospheric pressure to stop our blood boiling, and a fatal dose of solar radiation. These are just some of the challenges facing human space explorers, and we can also throw into the mix the apparent loss of gravity and issues with bone and muscle density, not to mention the psychological rigours of the journey. Fortunately, spacecraft have been developed to provide a life-supporting environment that allows astronauts to live and breathe in space, while advanced space suits carry further life-sustaining systems that allow an astronaut to venture beyond the spacecraft itself.

Providing artificial environments is sufficient for short-term excursions into Earth orbit or trips to the Moon and back, but one of the biggest challenges facing astronauts who spend long periods of time in space is the slow reduction in muscle and bone density because of the weightlessness. On Earth, the pull of gravity holds us firmly against its surface and we experience that as our weight. Without that gravity a simple jump would see us float off into space. Astronauts seem to display this effect of weightlessness but the gravity of the Earth is still very much present; in fact it is in some way responsible for the floating experience. Orbiting spacecraft are in effect constantly falling towards Earth, pulled by the force of gravity, and it is their forward motion that gives their path a curved trajectory which essentially matches the curvature of the Earth, preventing them from falling to the ground. Astronauts on board the International Space Station live in this weightless environment where they, the space station and everything inside it are falling at the same rate.

We have all experienced a time when we have momentarily weighed a little less, for example when driving too fast over a humpback bridge. As we, the car and everything inside it drops down a little faster than usual, our bodies weigh a tiny bit less for a fraction of a second, giving that stomach-in-the-mouth experience. You can take this to the next level and experience a 'zero G' flight – these take place on board large converted commercial airliners. After a conventional aircraft-style take-off you climb to a high altitude, then the pilot puts the aircraft into a dive. The aircraft must dive at a very specific rate so that everything inside falls at the same rate, and then for a few moments you are floating around inside the cabin. Just as it is for astronauts in orbit, gravity has not been switched off, you are simply falling at the same rate as everything else around you.

In this environment muscles and bones do not have to support body weight, so over time they weaken. Contrary to popular belief, an adult's bones are not solid unchanging lumps of calcium, they are very much an evolving part of the body, constantly reshaping and renewing themselves based upon the stresses and forces imposed on them. Studies have shown that long-term space exploration can reduce bone mass by as much as 1–2% per month spent in space. That is a significant weakening of the body's skeletal structure and it'll be particularly noticeable in the legs and lower back. We are all probably a little more familiar with the concept of muscle degradation, often caused by lack of exercise, and just like bones, the body's muscles will simply get weaker in a weightless environment. Of course this won't be a problem in space but as soon as you return to Earth you will notice the difference.

There are two ways to solve this problem. The first is to exercise, a lot. To that end, astronauts in space spend a lot of their time on special gym equipment. This is doable on journeys into space over many months to a year or so, but even then it is not possible to exercise for enough hours in the day to retain full strength. For longer-term space exploration, such as the mission we are planning, a more effective solution is to try and simulate the force of gravity on board the spacecraft. Contrary to what many sci-fi films depict, we do not yet have a magic device that when turned on causes gravity suddenly to appear. There have been experiments, however, using extremely powerful magnets which generate equally powerful magnetic fields, and one such experiment managed to levitate an unsuspecting mouse. The system was able to counteract Earth's gravity and was effectively producing a 1G (1G refers to the force of gravity that we feel on Earth) environment that suspended the mouse. Theoretically a similar magnetic system could be used in space to generate a 1G gravity field, but there is a problem with this approach. First we have no idea what impact such powerful magnetic fields would have on the human body, and second – and more practically – it takes phenomenal amounts of power to generate such huge fields.

There is, however, a more realistic solution that will help you on your journey. Instead of trying to generate a new gravity field, we can simulate the effects of gravity using other forces. You will have felt this already whenever you've been in a lift travelling upwards. As you stand in the lift and wait for it to start moving you will be experiencing 1G, but as the lift accelerates upwards, you will momentarily feel a little bit heavier. The acceleration of the lift means you get pinned against the floor a little bit harder, simulating a slightly higher pull of gravity. Just like this experience in an ascending lift, we could use linear acceleration to produce 1G. By constantly accelerating the spacecraft at the correct rate anything inside would be forced in the other direction, creating an experience similar to gravity, with the rear hull of the spacecraft becoming the floor upon which you could walk around as though you were on the surface of Earth.

Conventional technology using solid- or liquid-fuelled rockets is perfectly capable of simulating a 1G environment in this way, but the fuel would be used up in a matter of minutes. It might be possible to use a different fuel with a high specific impulse such as the VASIMR, although it is currently only capable of producing low levels of thrust. A more popular idea uses the concept of rotation, where inhabitants of a rotating spacecraft would experience gravity on the inside of the outer hull of the spacecraft. A large doughnut-shaped module could be rotated at a very specific speed to simulate gravity and the travellers on board could walk around and operate normally on the inside of the outer edge. Rotating any object at the right speed would cause simulated gravity on the outer portions of it. If you could rotate one of the rooms in your house at about seventeen revolutions per minute then you should be able to sit in a chair on the ceiling and read this book.

There are unwanted side effects of simulating gravity by rotation, and the main consideration in this respect is the Coriolis effect. You may well have heard of this before but it is a concept which is often misunderstood. Claims that it causes water to spin down a plughole in different directions in different hemispheres are wrong, but it does have an effect on our atmosphere. A great example is the movement of a parcel of air across the surface of the Earth in the northern hemisphere. Let us consider such a parcel moving from the polar regions towards the equator: it does not follow a straight line with respect to a weather watcher on the ground; instead, the rotation of the Earth causes it to turn in a clockwise direction with reference to its direction of travel, and it is this effect which gives us the rotational nature of high and low pressure systems. The apparent force from the Coriolis effect acts at right angles to the rotation axis, so an astronaut moving towards or away from the rotational axis would experience a force pushing towards or away from the direction of the spin. This would lead to feelings of dizziness and nausea, and the only way to overcome these would be to reduce the rate of spin to lower than about two revolutions per minute, but that would require a much larger rotating doughnut to produce the required 1G effect.

If we could simulate gravity then muscle and bone mass loss would be minimized if not eradicated and the whole experience of a long-duration flight around the Solar System would be much more agreeable.

The psychological challenges are huge, not the least among them being the necessity of getting a decent night's sleep on board. The varying levels of lighting on board the International Space Station make it difficult for the mind to hook into night and day because the usual cues to help regulate the body clock – a blue sky, or a daily sunrise – are just not there. In fact astronauts on board the ISS see about fifteen sunrises and sunsets every day. Even something as simple as looking out of the window before sleep can send the wrong signals to the brain, leading to a disturbed night's sleep.

Probably the biggest psychological issue to overcome, however, will be the sense of isolation – and that's regardless of whether you are on your own or not. Humans are by nature sociable creatures so to be separated from the rest of mankind will take its toll. Studies on astronauts have shown that the longer the mission, the greater the feeling of isolation. This effect is worsened when the Earth is out of view, as experienced by Apollo astronauts when they were on the far side of the Moon. This is known as the 'Earth-out-of-view' problem, and it will only become more of an issue as you travel further into the Solar System.

This anticipated isolation is just one of the reasons why astronauts must pass a series of psychological assessments, as well as character and personality tests. One of the most important factors for consideration when selecting astronauts is for them to have 'the right stuff', to be calm and rational under stress and great at working in a team as well as when alone. Certainly in order to undergo this mission you will need to be strong in body and mind.

Preparing psychologically and physically for your epic journey through the Solar System is therefore essential. Whatever the mode of transport in space, you first have to get there, and to do that you need to go fast – about 11 kilometres per second to escape Earth's gravity. On launch, you will experience a force of 3G, which is three times the force of gravity experienced at the surface of the Earth. Some fairground rides will exert this kind of force on the body but a better training approach is to spend some time on a human centrifuge. These installations are used by professional astronauts and simply comprise a large device with a long arm which has a capsule at the end. You sit inside the capsule and the whole thing rotates around at speed. The device is capable of producing as much as 20G but astronauts experience only about 3G. Fighter pilots can experience higher G-forces for shorter periods of time. With a special G-suit on, a trained person can sustain about 9G, maybe 10G, but not much more. So with a human centrifuge the launch experience can be simulated and it is possible to prepare for the sensation of taking off and the physical experience of space flight itself.

As I am sure you have gleaned from this chapter, there are many challenges facing you as you embark on your journey around the Solar System. Astronauts have even reported missing seeing colours on long-term postings to the International Space Station, so even that aspect needs to be considered. And we haven't even touched on the issues of food and hygiene for the journey, although we will do so along the way.

It seems, then, that our voyage is set. The spacecraft will be launched into Earth orbit on board a conventionally fuelled liquid-/solid-fuelled rocket before using an ion engine for low, long-term thrust. Your craft will be called the *Kaldi* after the Ethiopian goat herder who is said to have discovered coffee – the other fuel that will keep you going on your long journey. It will use VASIMR to power it into interstellar space and it will rotate at less than two revolutions per minute to provide a simulated 1G environment. Finally, so that you can take a proper look at the planets of the Solar System as you visit them, we will have to momentarily forget about the laws of physics and employ a device called the Reality Suspension Unit, or RSU. This will enable you to get up close to the planets and explore them in all their glorious detail. Which is, after all, one of the main purposes of your mission.

TWO

Goodbye Earth

DEPARTURE FROM EARTH IS an emotional experience. As you stand at the foot of the Titan rocket, about to board, it dawns on you that the next breaths you take are the final few times fresh air will fill your lungs. You turn round to take one last look at the surface of Earth. All of a sudden the months you have spent training and preparing for this moment seem completely inadequate.

With your movement inhibited by the pressure space suit, which is an essential safety element of every launch, you find it a struggle to mount the steps then bend to get through the Titan's tiny door. Inside, the conditions are very cramped. Assistants ensure you are connected up to the comms system and the life support system in case of depressurization on the way up. More importantly, they make double sure you are strapped in, because it's going to be a pretty bumpy ride. At last the door closes and there is silence, except for the gentle hiss of the life support system and whirring and clicking from other equipment. Your epic journey has begun. At this moment, you feel more isolated than ever before.

The rocket engines fire. A thunderous noise fills the air and the cabin starts to shake violently. Eventually the words crackle through the intercom: 'We have lift-off.' You can feel it. The acceleration is incredible. You feel yourself being pressed hard against the seat: you are pulling about 3Gs on launch. With the noise and the shaking your senses are on full alert as they feel like they are being attacked on every level.

Most of the launch is automated, but there are one or two simple but critical tasks for you to perform, such as initiating the controls to separate the various stages of the rocket, each bringing with it a rather unnerving jolt. Almost as suddenly as it all started, though, you fall into silence as you reach Earth orbit. You feel strange, as though you are going over a humpback bridge but never landing. Weightlessness is going to take some getting used to. You unbuckle your belt and float almost gracefully out of your seat. This is really it!

Although for the launch itself you were required to wear the pressurized space suit, this cumbersome clothing can now be taken off and replaced with a more comfortable jumpsuit. Before long you can see out of the viewing window the International Space Station, with which you must dock to fill up with supplies: it is more efficient to launch with minimal payload and then top up somewhere else. After a brief stop, the *Kaldi* undocks and slips away; its living quarters start to spin up slowly and silently restore gravity, and a sense of normality begins to pervade life on board.

You glance out of the window to take a lingering look at Earth before departing on the long voyage. Seeing your home from space with such clarity you notice just how beautiful and vulnerable it appears. It looks perfectly circular but in reality Earth is an oblate spheroid, which means it is very slightly squashed at the North and South Pole as though being gently squeezed between finger and thumb. The diameter across the equator is 12,756 kilometres but measured through the poles it's 41 kilometres shorter at 12,715 kilometres. Earth's equatorial bulge is present because of its rotation, which we perceive as the rising and setting of the Sun, Moon, planets and stars. It is this rotation which gives us a measure of one day as twenty-four hours, although in reality it takes the Earth 23 hours, 56 minutes and 4 seconds to complete one rotation, and we just round it up. That discrepancy of 3 minutes and 56 seconds means that every day the stars seem to rise 3 minutes and 56 seconds earlier and slowly creep further towards the western horizon from day to day. You will remember from the last chapter that artificial gravity can be created by rotation. As the Earth rotates, material around its equator is forced into a curved path, but there is a resistance to this movement which results in the equatorial bulge.

The other thing you can see from your viewing window is the white light of the Sun. From Earth it always looks yellow, or possibly an orangey/red when lower down in the sky, but we can thank the Earth's atmosphere for that illusion. The gas molecules that make up our atmosphere scatter light at the shorter end of the wavelength, which is why the sky looks blue, but the other end, around yellow, orange and red, remains less affected so the Sun appears more yellow than it really is. When viewed from space with optical protection and without the effects of the atmosphere, you can see it in its true colour, white.

As your first orbit of Earth continues, the Sun slowly drifts behind the Earth and you slip into our planet's shadow. Another incredible sight appears as you see the stars, their crisp, pure light unaffected by the 100-kilometre-thick blanket of gas that surrounds our planet. With the Sun out of the way they appear brighter and clearer than you have ever seen them before. The atmosphere of Earth is constantly on the move as the energy of the Sun heats it, making it rise and fall. The rising and falling air creates pockets of low and high pressure which make the air move around in a horizontal fashion, producing wind. All of this movement – and this is a very simplistic view of an incredibly complex system – causes the incoming light from distant stars to bounce around and get disturbed. This, plus dust and pollutants in the atmosphere and of course the ever-present cloud, means the appearance of the stars is often distorted. From up here in orbit, though, they look incredible.

Orbiting around the Earth at an altitude of about 400 kilometres, the continents and oceans come and go. Something that is less evident is the tilt of the Earth's axis of rotation which has a very important part to play in our weather system. All of the planets, the Earth included, orbit around the Sun broadly in the same plane. This plane can be seen in the night sky as the Sun, Moon and planets all roughly follow the same path. You can imagine this by thinking of the Sun sitting at the centre of a giant sheet of heat-resistant paper and the planets orbiting it. As they spin, they all, with the exception of Uranus, tend to sit roughly upright with reference to the plane of the Solar System. They do all exhibit some tilt, however, and in the case of Earth the tilt is about 23.5 degrees to the vertical. This tilt is directly responsible for the seasons we experience. When the North Pole of Earth is tilted towards the Sun, its heat has to travel through less atmosphere in the northern hemisphere so we experience the warmer weather of the summer. When the North Pole is pointing away from the Sun then its energy has to pass through more atmosphere before hitting the surface so we experience the cooler weather of the winter. While northern hemisphere dwellers are experiencing summer, the southern hemisphere is pointing away from the Sun and experiencing winter.

The direction in space in which the axis of rotation points is known as the north celestial pole, and from Earth that point happens to lie very close to the star known as Polaris. But much like a spinning top, Earth is wobbling in space, completing one wobble every 26,000 years, so in a few thousand years' time, Polaris will no longer be the North Pole Star.

Above the North Pole of the Earth it is possible to see an eerie auroral display in progress. This beautiful natural phenomenon is the result of the interaction between electrically charged particles from the Sun, the solar wind, and the atoms of gas in our atmosphere. The solar wind is a fairly steady stream of charged particles that emanate from the Sun but on occasion there is a burst of higher intensity from coronal mass ejections. Depending on where the wind has come from on the Sun, it will travel at either 400 or 750 kilometres per second so it takes a day or two for the outbursts to be felt on Earth. On arrival, they are channelled around the planet's magnetic field, accelerating particles already within the field to higher speeds which then drop down into the denser regions of the atmosphere around the North and South Poles.

What happens next requires a little knowledge of atomic physics. We have already seen that atoms are made up of a nucleus containing protons and neutrons surrounded by a number of electrons in orbit. The electron orbits are at very specific distances from the nucleus and, if at all possible, the electrons like to sit in their rightful place. By giving the electrons some energy, they move into higher orbits (further from the nucleus), and as soon as they can, they dump the energy as light and drop back to their usual orbit. The light dumped by the electrons is what we see as the glow of light from the aurora borealis (known as aurora australis in the southern hemisphere). The displays are usually easy to see around the polar regions but less common the further away from the poles you are. If there is a particularly big burst of solar material and the conditions are right, then auroral displays can be seen at lower latitudes. They are an astonishing sight.

It is not uncommon for astronauts to see meteors as they plummet through the Earth's atmosphere, although there is an element of luck in spotting them, as is the case from the Earth's surface. Seeing a meteor from space must be a really graphic reminder of the dangers of space flight. These tiny pieces of rock, which usually measure only a few centimetres across, travel through space at speeds in excess of 90 kilometres per second. That is significantly faster than the fastest bullet, which travels at a mere 1.4 kilometres per second, so imagine one of those hitting a spacecraft, or worse, hitting you when you are out on a spacewalk.

From Earth, these interplanetary pebble-sized chunks of rock usually become visible to us as they hit the upper layers of our atmosphere at an altitude of about 100 kilometres. As they fall they crash into atoms of gas, and the high-speed impact dislodges material from the meteor in a process known as ablation. The disturbed gas atoms momentarily have electrons stripped from them which produces a trail of positively charged atoms and negatively charged electrons. After a short while they recombine and give off a little light which we see as a bright trail behind the meteor – hence the name 'shooting star'. It is not uncommon for larger meteors to plunge further into the atmosphere and compress the air so much that a shockwave builds up, producing a sonic boom. All but the largest meteors will completely disintegrate by the time they reach an altitude of about 50 kilometres. Those that do make it to the ground are known as meteorites and are surprisingly cool to the touch despite their violent journey through the atmosphere.

It's a rather more worrying phenomenon up where you are. Meteors pose a real threat when it comes to space travel. There have been a few reports of impacts on the International Space Station and other spacecraft, but to date no one has been killed. Back in 2012, an impact event occurred to a viewing window in the Cupola module of the ISS. Just like a stone or piece of grit will chip the windscreen of your car, so will tiny pieces of space rock chip windows on spacecraft. In the case of the ISS the windows are made of fused silica and a material known as borosilicate which makes them literally bulletproof. Most modern spacecraft are now made from a very special multi-layered shell which includes a Kevlar-style woven fabric to reduce the velocity of any potential impactors. Astronauts outside this protective shell are vulnerable, however, and should a strike occur, the results would undoubtedly be fatal.

Time is of the essence on this mission so you cannot loiter around in Earth orbit any longer than necessary. It is now time to fire up the booster engines to take you to your first port of call, the Moon. Not firing the engines at the critical point will mean that you will more than likely arrive at the rest of the planets too late and they will have moved further along their orbit. Launch and rocket burn windows are more critical with chemical rockets as it takes a much greater amount of fuel to adjust trajectory compared to spacecraft powered by the VASIMR system. If we were to use that instead, it would lead to an orbit of Earth that would slowly spiral out towards the Moon. A similar system was used on the SMART-1 mission and from an orbit of just a few hundred kilometres it needed to increase it by about 384,400 kilometres – the distance of the orbit of the Moon. At the low thrust of the electric engine it took a little over a year to get there but it only used about 100 kilograms of fuel. It is an interesting contrast to the Apollo 11 mission which used almost 6,500 kilograms to get there in a little over three days. Clearly electric engines are significantly more efficient.

The coast to the Moon gives you plenty of time to get used to the experience of living in space. Weightlessness is a strange sensation that takes quite a few weeks to get used to but spinning up the habitable portion of the ship reinstates 'gravity' again. The rooms are arranged in a linear fashion around the doughnut-shaped portion of the craft so that the floor is actually the outer edge of the hull. There are sleeping quarters, living quarters, a kitchen, storage rooms, and even a bathroom.

The voyage will be very comfortable, but without the luxury of artificial gravity things would be very different. On a normal space flight vacuum power is used heavily to control the movement of things on board. It's particularly useful for maintaining personal hygiene: for example, the toilets will use a vacuum flush system instead of water. Some toilets have handles to help the astronaut sit tight; some even have leg and foot restraints. Solid waste is simply sucked down into the toilet by rotating fans that produce suction through a tube connected to the bowl of the toilet. The stools are distributed into storage containers that are exposed to a vacuum to dry them before eventual transportation back to Earth. The vacuum tube also has a detachable urine receptacle with male and female adaptors, although unlike the solid waste the urine is collected and then dumped overboard. Interestingly, after launch, the lack of gravity causes all bodily fluids to spread evenly around the body. This movement is detected by the kidney which triggers a physiological reaction: astronauts must relieve themselves within a couple of hours of leaving Earth. For this reason there is a rigid diet and bowel movement regime in place for all launches. Thankfully, simulated gravity means you can enjoy normal bathroom activities for the duration of the flight.

The rooms on board the *Kaldi* are pretty similar to the ones back at home. The only difference is that they are not all on the same level: the rotating nature of the craft gives the sense of gravity on the inside of the outer edge so it is this face which acts as the 'floor'. But all the comforts of home are present, from a cosy bed to a sitting area and flowing water for cleaning and cooking. What gives the rooms a surreal look is the pitch black outside the windows twenty-four hours a day, 365 days a year. All of this is in quite stark contrast to your recent experience on board the International Space Station which had no gravity, and where water did not flow in the conventional sense and personal hygiene tasks were much more difficult to carry out.

For the first day en route to the Moon it looks no different to how you are used to seeing it from home, although it might seem more prominent against the inky blackness of space that is now no longer dulled by looking through the Earth's atmosphere. The Sun, Moon and Earth are now all visible and they seem eerily similar to the diagrams you see in textbooks showing how their changing relative positions cause the appearance of the phases of the Moon. The Earth lies behind you, the Moon roughly ahead, and the Sun is off to the left or port side of the craft. In that configuration, the

Moon would appear to those on Earth as a quarter Moon with half of the visible portion being illuminated and the other half in darkness. The line between the two where day meets night is known as the terminator, and the detail along this line always looks stunning. It is here that the shadows are at their longest, making surface features stand out.

Of all the surface features on the Moon probably the most well known are the craters and vast grey plains known as the seas, or maria (the plural of 'mare', the same as the Latin word for 'sea'). Before the invention of the telescope it was generally believed that these grey patches were large bodies of water, but closer study (chiefly by Galileo) revealed that they are simply vast plains darker than the surrounding areas. The impacts of chunks of rock are to blame for the landscape of the lunar surface, with craters formed by smaller impacts and bigger collisions creating the large plains. When the Solar System was younger there were large pieces of rock left over from its creation and when they hit the Moon they cracked the crust and let molten lava flow through from the mantle below. This lava filled the impact crater and, over time, solidified leaving behind the plains we can see today. It is possible to roughly date the age of the lunar surface by studying the size and distribution of the craters. A very new feature, for example, is less likely to have been disturbed by subsequent impacts than an older feature.

You'll enjoy seeing the familiar features of the Moon as you fly around it for the first time. Perhaps the most famous is the Sea of Tranquillity, where Armstrong and Aldrin took those historic footsteps. From here, up in space and without the distorting effects of the Earth's atmosphere, it appears clearer now than ever before, somehow closer, as though you could reach out and touch it.

Then you will get to see something very few humans have seen, the far side. The concept of the far side of the Moon is a strange one but it is not uncommon among other natural satellites in the Solar System to have a hemisphere that remains forever turned away from the planet. Many people believe that the Moon's far side is in permanent darkness, but in fact at some time or other the entire lunar surface experiences sunlight. It rotates once on its axis every 27.3 Earth days so common sense suggests that if you were standing on one spot on the surface of the Moon, you would see a sunrise every 27.3 days. Rather confusingly, you would actually see a sunrise every 29.5 days – a discrepancy created by its orbit around the Earth. In the time it has taken the Moon to spin once on its axis, the Earth has moved around the Sun a little, dragging the Moon with it. This means the Moon has to spin a little bit more – for 2.2 days as it turns out – to account for its slightly different position in space.

It isn't a cosmic coincidence that the Moon takes 27.3 days to rotate once on its axis and also 27.3 days to complete one orbit of the Earth, and this is known as captured or synchronous rotation. Behind the scenes it is gravity that's responsible for locking the rotational and orbital periods of the Moon, and it's all linked with the effect we know as the tides.

The Moon, like all objects, exerts a gravitational pull on the Earth and that pull produces a slight bulge on the side of the Earth facing the Moon. As Earth rotates, any locations passing through this gravitational field will experience a rise in water levels, and to a much lesser extent a rise in land levels too. The rising water levels are what we know as high tides and they can be seen at nearly every location on the globe at some point: there are places on Earth where tides do not occur, for example at the North and South Poles at certain times. For the rest of the planet, there is another tidal bulge.

Because the strength of gravity decreases with distance, another bulge is produced on the opposite side of the Earth. Effectively the material inside the Earth experiences a greater pull than the surface material (such as water) on the far side so it is more accurate to say that the Earth is being pulled away from the oceans, creating the second but slightly lower high tide roughly twelve hours later.

Crucially, the main bulge on the lunar side does not sit exactly on a line between the two objects. As the Earth spins, it drags the bulge with it, causing it to lie a little ahead of the line between the Earth and the Moon, and this misalignment of the tidal bulge means that the extra mass of the bulge results in an extra pull of gravity which tugs on the Moon making it accelerate in its orbit. This acceleration causes it to move further away from the Earth at a rate of about 3.8 centimetres per year.

With the Moon and Earth locked in an orbital dance around their common centre of gravity, known as the barycentre, it is perhaps reasonable to expect that on occasion they will block one another from view. For example, we see an eclipse of the Moon when the Earth lies directly between the Sun and Moon. It is also this rough alignment that gives us the full Moon phase, yet we do not see a lunar eclipse every time there is a full Moon. This is because the orbit of the Moon around the Earth is tilted by about 5 degrees with respect to the orbit of the Earth around the Sun so the three of them are not always in a perfect line. This means at most full Moon phases that the Moon will lie just above or just below the Earth–Sun line so sunlight will not be blocked by the Earth. Similarly we see a solar eclipse when the Moon blocks our view of the Sun, which only happens at new Moon, but on most occasions the Moon is either a little above or a little below the Sun. A perfect alignment of three celestial bodies like this is known as a syzygy. Solar eclipses are particularly spectacular because the Moon and Sun are both roughly the same size in the sky so usually the presence of the Moon in front of the Sun means it blocks out the Sun's bright photosphere, revealing its beautiful yet faint outer atmosphere, known as the corona. There are also partial eclipses where either the Moon is only partly in the Earth's shadow or it only partly blocks the Sun from view. And then there is the annular eclipse where the Moon is very slightly smaller than the Sun in the sky, which is a result of the elliptical nature of the Moon's orbit around the Earth. When the Moon is at its furthest point from the Earth it appears at its smallest in the sky, and during an eclipse a ring (or 'annulus') of the Sun's photosphere is visible behind it.

On arrival at the Moon there may be a need to execute a long fire of the VASIMR engine to adjust the trajectory and send you on to the first planet on your journey. Firing engines on arrival (or close to arrival) at destinations is not unusual, whether it is to adjust the path of the spacecraft a little so that it follows the necessary trajectory to slingshot correctly on to its next destination, or to slow the spacecraft so that it can drop into orbit. If the velocity of an approaching spacecraft is too high then it will simply slingshot around. If the velocity is too low then it will crash into the surface, so getting it right is essential.

This was beautifully demonstrated in 2011 by the Gravity Recovery and Interior Laboratory spacecraft which was composed of two lunar orbiters. During a carefully choreographed four-month slow cruise to the Moon, the two spacecraft managed to achieve a separation of about 100 miles which was crucial to the success of the mission. Their purpose was to map the interior structure of the Moon by monitoring their relative positions. The two spacecraft were equipped with a system that used electromagnetic waves to tell them exactly how far apart they were and how far away from Earth they were too, to an accuracy of a hundredth of the width of a human hair – equivalent to about 1 micron. The systems on board emitted electromagnetic waves whose frequencies were monitored from Earth and by each other. As they tracked above the lunar surface at an altitude of about 23 kilometres, the frequencies monitored shifted as the spacecraft were tugged downward and accelerated or drifted up and slowed down. The changing speed was the result of a changing gravitational field caused by 'heavier' material below them in the interior of the Moon. By measuring the changing speed it was possible to weigh the Moon at that location and therefore build a picture of its internal structure. After many orbits over the four-month period a gravity map of the entire lunar surface was completed, giving an unprecedented view of the inside of the Moon. Other missions have helped to develop this knowledge. For example, magnetometers have been carried to the Moon on various missions and are used to measure the magnetic field at different locations, while lunar landers have studied rock compositions. By piecing all this information together we now have a good understanding of the internal structure of our nearest astronomical neighbour – and that can help tremendously when planning missions.

We have learned that the Moon has a solid iron-rich inner core with a diameter of 480 kilometres which is surrounded by a liquid iron outer core that has a diameter of about 610 kilometres. Surrounding the outer core is a partially molten boundary layer which is an estimated 200 kilometres thick and which separates the core from the mantle and crust. The formation of these outer layers is thought to have followed a period of fractional crystallization of a magma ocean that covered the entire lunar surface. This process is one of the most important in the evolution of planetary geology. Silicates were removed from the molten magma, and eventually about three-quarters of the magma crystallized, creating a mantle rich in magnesium, iron, pyroxene and olivine while lower-density materials such as plagioclase and anorthosite (minerals of the feldspar family, the latter igneous in nature, having formed from the cooling of magma) rose to the surface forming the crust that we see today, which is thought to be about 50 kilometres thick.

The fractional crystallization process in the magma ocean is strong evidence that the Moon had a violent birth. A very large object about the size of Mars is thought to have struck Earth around 4.5 billion years ago and sent debris flying out into space. Over millions of years, the material suspended in orbit around Earth coalesced into the Moon we see today. The vast amounts of energy involved in the impact meant that large portions of the Moon were liquidized, leading to the formation of the magma ocean.

Thanks to other orbiting spacecraft such as the Lunar Reconnaissance Orbiter (LRO), which used laser altimeters, we now also have a huge amount of detail about the surface topography. Cameras on board were even able to take pictures of some of the Apollo lunar landers which still sit dormant on the surface. Such topographical maps were not available for the Apollo missions so NASA used surface drawings created by Sir Patrick Moore for their Moon landings, such was the quality of his work. Because of orbiters like LRO we not only have detailed maps of the near side of the Moon but of the far side features too, and there are quite significant differences between them. The lunar maria you looked at earlier are found almost entirely on the near side of the Moon and account for just over 30% of the surface features; only 2% of the far side is covered in maria. It is thought that a thinner crust and a concentration of heat-producing elements under the crust on the near side are responsible for this difference. As the Moon formed and the crust solidified, elements that are incompatible and were separated in the liquid became trapped together in the region between the crust and mantle. Uranium and thorium are just two of these elements, and their radioactive properties produced a lot of heat. The heating caused a partial melting of the upper layers of the mantle which were subsequently brought to the surface through volcanic eruption, or through cracks and fissures in the crust from meteoric impacts.

The remainder of the lunar surface is made up of the highlands, named simply because the surface level is generally higher than the plains, not because of rolling hillside and mountain chains. As we know, the craters are formed by meteoric impact, and if you have ever looked at the Moon through a telescope you will know there are thousands of them. Estimates suggest there are just over a quarter of a million with a diameter of a kilometre or more, and that's only on the near side; there's at least the same number on the far side. There are some craters near the polar regions on the Moon whose bases are almost permanently in shadow, and it is in these craters that water ice has been discovered.

Water in its liquid form cannot exist on the surface of the Moon chiefly because of the exposure to solar radiation. Liquid water on Earth is protected from such an onslaught, but when exposed to the radiation on the Moon, water molecules soon decompose in a process known as photodissociation – essentially a chemical reaction that breaks down the hydrogen and oxygen in water. It has long been thought that water in its ice state may well exist in deep craters and in the sub-surface layers, having been deposited on the Moon by cometary impacts. In 1998, an instrument known as the neutron spectrometer on board the Lunar Prospector spacecraft revealed high concentrations of hydrogen in the top metre of the lunar surface, indicating the presence of sub-surface water, similar to permafrost found in regions on Earth. Study of rocks returned by the Apollo missions also revealed evidence of water, so it seems that water is a common commodity on the Moon. This has since been confirmed by the Chandrayaan-1 spacecraft, which in 2008 found conclusive evidence of water ice in polar craters that received no sunlight. This is all great news for potential lunar habitation.

You will find that the craters on the Moon look stunning close up. Even through a telescope from Earth it is amazing just how much detail can be seen. It's surprising, too, to see how different they can all appear, from the single young craters that have pierced the pristine lava plain of the maria and the old craters that have almost been obliterated by newer impacts through to the complex crater chains seen stretching across the surface. We have already touched on the cause of craters and the fact that on Earth the atmosphere acts as a protective shield against rock impacts. That is why our planet is so devoid of craters – that and the constant erosive effects of the weather, which over the centuries gently erase signs of impact. Although there are a few craters left on Earth like the majestic and aptly named Meteor Crater in Arizona, on the Moon the lack of weathering from an atmosphere means that craters remain for millennia, unless another impact obliterates them. Even the footprints of the twelve Apollo astronauts who have walked on the surface remain for all to see, and will do so for millions of years to come.

One of the most prominent craters is Tycho in the southern lunar highland region, named after Johannes Kepler's old tutor, the Danish astronomer Tycho Brahe. Not only is it an almost perfect example, it also exhibits two quite remarkable features. The first is particularly noticeable when the Moon is full, which is not the usual time to make lunar observations: usually the features look best when the Sun is low so they appear more prominent when along the terminator. At full Moon, Tycho's stunning ray system bursts into view, stretching for 1,500 kilometres across the surface and centred on the crater itself, which measures 86 kilometres in diameter. Theories about the origin of ray systems have varied over the years but have included salt deposits from water through to deposits of volcanic ash. What we now know is that their nature is intrinsically linked to the formation of the craters themselves. As the impactor strikes the surface, the energy released blasts quantities of the lunar regolith (the unconsolidated layer that sits on the bedrock) and maybe even fragments of the impactor itself outwards from the impact site. This material then settles back on to the surface as the ray systems that we can see, and it is clear from their bright colour that they tend to be from relatively newly formed craters. In the case of Tycho, that's about 108 million years old – young in geological terms.

Another very common crater feature which you'll notice is a central almost mountain-like peak. They are more difficult to observe than the ray systems and are best seen when the crater sits on the terminator. For Tycho, this means when the Moon is about nine days old as the incoming light from the Sun casts long and deep shadows on the floor of the crater so the peak looks quite prominent. The origin of the peaks found inside craters is known to be the result of the energy involved in the impact when the crater forms. As the rock slams into the surface of the Moon the release of energy is sufficient to partially melt the surface material. The force of the impact sends a shockwave outward which rebounds back into the centre, taking with it some of the molten material which then forms into a peak as it solidifies.

Mare Crisium is a great example of how craters and maria are linked. If you look around the edge of Crisium you'll see that it has a very jagged appearance and is circular in shape. Clearly this was once a crater that has since been filled by molten lava. Across the surface of the Moon you can see other examples of curved jagged mountains which are actually ancient crater walls that have since been modified by more impacts. When looking at the Moon it is actually quite hard to find much undisturbed material, but there are some areas where it is possible to see what the surface is really like.

As you swing past the Moon, this is your first chance to try out the Reality Suspension Unit, which will allow you to pop down to the surface and have a walk around. 'The surface of the Moon is fine and powdery' are among the many words uttered by the first human visitors, Neil Armstrong and Buzz Aldrin, and those words will come to mind now as you take your first steps. Walking on the Moon for the first time is an incredible experience and the view from the surface really is that of an alien world. It is as far removed from the views we get on Earth as we can possibly imagine. Unlike Earth, which is teeming with life, on the Moon there is no sign of it, anywhere. It is a completely desolate and inhospitable world. It is a humbling and emotional experience to be among the first living creatures to set foot on this landscape since it formed over 4 billion years ago.

The surface that Armstrong and Aldrin described, and which you are now walking on, is known as the lunar regolith and consists of countless pieces of tiny particles the same size as sand and silt. The particles have been created over billions of years by a bombardment from rocks that pulverized the surface material over and over again, much like a chef with a pestle pounds whole spices into a powder in a mortar. As you might expect, there are larger chunks mixed in with the finer particles, but overall it has a powdery feel. The material covers the Moon's entire surface to depths ranging from 5 to 15 metres, but it gets quite compacted just a few inches down. Armstrong even reported that it had almost snowlike properties, sticking to their boots as they moved around. The Moon's regolith is not unique in the Solar System, though, as we shall see. A fine powdery surface exists on Mars too, as well as a number of asteroids and even a few other planetary moons.

Something else that is making your Moon walk quite an alien experience is the noticeable absence of atmospheric perspective. If you look at a scene on Earth, perhaps a range of mountains in the distance, they appear faded and almost a little more blue than your immediate surroundings. This fading effect is caused by dust in the atmosphere which tends to block some of the light coming from that distant scene; it's one of the key ways in which our brain is able to determine that things are far away. The contrast too seems to reduce over distance, bringing those distant objects closer to the apparent colour of the sky. The presence of the dust particles in the atmosphere means that some of the already scattered blue light through the atmosphere gets reflected back towards you, the observer. Having an atmosphere that can suspend dust in it is essential for this perspective effect to occur. On the Moon there is very little atmosphere. In fact it is so rarefied it is almost a vacuum, and its virtual absence means no suspension of tiny particles and therefore no atmospheric perspective. It is for this reason that it is very difficult to judge distances to lunar landmarks such as craters and mountains, which always seem a lot closer than they really are.

The lack of an atmosphere on the surface of the Moon would make it a wonderful base from which to carry out astronomy. The Earth's atmosphere has a number of effects that thwart astronomers in their attempts to get sharp images of the night sky. The most noticeable is known as scintillation. You are already familiar with it as the twinkling of stars in the night sky. As light from a distant object passes through the atmosphere it gets deflected from its path, making the object seem to jump around. If you have ever looked through a telescope then you will already know that the image of a distant object can be spoiled terribly when the atmosphere is particularly turbulent. This is one of the reasons why professional observatories are often placed at the top of high mountains, above as much atmosphere as possible. The atmosphere also blocks out some of the wavelengths of the electromagnetic spectrum coming from distant objects so we are unable to study them from the surface of the Earth. This has a profound effect on our ability to fully understand the nature of objects in the Universe since we must study them in all wavelengths of light. A great analogy is the sound coming from an orchestra: to fully appreciate it we must listen to the music from all the instruments, not just one or two of them. There are ways to counteract the effects of scintillation but some level of detail will always be lost. We have come to accept that the blocking effects of the atmosphere are not something that can be worked around, which is why astronomers have taken to launching telescopes into space.

The Moon would be an ideal alternative to this expensive and challenging operating environment. Its lack of atmosphere means that even ground-based telescopes would get fantastic views. The fine dust might be a problem but there is very little movement of the regolith material because there is no wind. Astronomy could even be done during daylight hours because the sky is black not blue during the lunar day. There are craters near the poles of the Moon whose bases are nearly constantly in the dark and therefore very cold, making them ideal locations for infrared telescopes, and the far side of the Moon would be great for radio telescopes as they would be protected from the constant chatter of noise from our terrestrial radio transmissions.

Like certain craters whose floors remain permanently in the dark, there are some mountain peaks around the Peary Crater at the Moon's North Pole that remain constantly illuminated. These are known rather romantically as Peaks of Eternal Light, and there are other such places on other bodies in the Solar System. Not all planets have these points though. For areas on a body to remain permanently lit they must be high altitude, and if the axial tilt is small then they must be near the polar region. The tilt of the axis of the rotation of the Moon is small, at just 1.5 degrees, but Earth's is over 20 degrees so there are no regions like this on Earth.

The low surface gravity on the Moon also makes it a great base camp for an exploration of the outer planets. To launch a spacecraft into space from Earth requires a speed of 11.2 kilometres per second, and in the case of the Apollo missions this required 5.6 million pounds of fuel, but the return trip was far more efficient because of the Moon's lower gravity. The escape velocity of the Moon is just 2.4 kilometres per second so the Apollo missions required just 40,000 pounds of fuel to take off and return home. Add into the mix that there is a great supply of liquid fuel available on the Moon in the form of water and it makes a lot more economic and logistical sense to explore the Solar System from the Moon.

Fortunately for you, though, your lunar excursion can be brought to an end simply by returning to the ship. Back on board, your trajectory is adjusted very slightly by the gravity of the Moon and you swing around on course for an encounter with the Sun, the star that is the powerhouse of our Solar System.

THREE

Into the Furnace

As if hurtling off into the depths of space in a tin can attached to a rocket was not dangerous enough, the next phase of the journey has to be the most perilous yet. The leg of the journey that takes you past the Sun will subject the ship to immense heat and radiation the like of which has never been experienced before. Earth's magnetic field, and to a lesser extent its atmosphere, protects us from harmful radiation from space, the majority of which comes from the Sun. Getting close up is a risky business.

It isn't just solar radiation that's a problem though. Astronauts have often reported seeing flashes of white which have been attributed to cosmic rays passing through their heads. A scary thought indeed. Clearly long-term space flight away from the protection of the magnetic field of Earth requires serious consideration. One new solution that has been incorporated into the structure of the *Kaldi* is superconducting magnets to reproduce the magnetic field of the Earth and thus protect those on board from radiation. The principle is pretty simple and it relies on the fact that a superconducting material has zero resistance so that a large electrical current can be easily passed through it, the flow of which produces a magnetic field. With the magnetic field in place, radiation such as solar and cosmic rays should not be able to penetrate through to you inside.

On the approach to the Sun the temperatures are going to increase quite significantly. To protect the *Kaldi* from the heat of the Sun's infrared rays, sensitive areas are wrapped in gold foil. Material like this reflects the infrared radiation and has been used on spacecraft for some time. The windows too have sliding shutters made of a polycarbonate and coated in a thin film of gold, much like astronaut helmet visors. Gold is the material of choice because it has an ability to stop infrared radiation getting into your eyes. At the back of the eyeball are light receptors known as rods and cones but these only detect visible light. If too much visible light is received then the brain instructs the muscles to close the eyelids or turn the head away, but with no infrared detectors the brain does not know if too much infrared energy is hitting the retina, so no such response is triggered and it is very easy for the retina to get burned, which can lead to blindness. Even behind the protective layers of the Earth's atmosphere it is dangerous to look directly at the Sun. In space, the gold film on visors offers protection against infrared but they are still vulnerable to visible light because complete protection would make the visors far too dark for astronauts to see what they are doing. Looking at the Sun is a real danger in space, especially when venturing outside the spacecraft, and astronauts need to be particularly aware of where the Sun is at all times.

Fortunately, your craft has extra protection in the form of an experimental system made of Mylar. This is a material that is familiar to amateur astronomers who use the stuff, which somewhat resembles thin aluminium foil but is actually a polyester film coated with a very thin layer of aluminium that blocks out 99% of the visible light from the Sun. By placing it over the front end of a telescope, amateur astronomers can safely and directly observe magnified views of the Sun. Placing Mylar over one set of window blinds means you can simply close that blind and safely look at the Sun as you approach.

Visually the Sun looks no different, although it's obviously a little bigger by now. The clarity of sunspots is pretty staggering though, even without any magnification. Sunspots, as their name suggests, appear as spots on the surface of the Sun, or more accurately on the visible surface of the Sun, the photosphere. The Sun is a huge ball of plasma, the fourth state of matter, and the plasma making up the Sun is essentially gas in the same state as the stuff in the VASIMR engine, although the solar variety is composed of different elements – primarily hydrogen, a little helium and traces of other elements. By looking (safely) at the Sun, you are actually just looking at the top of the visible surface of plasma. There is no surface upon which you could land.

Even from Earth, some 150 million kilometres away, we can learn an awful lot about the interior of the Sun. Indeed nearly everything we know has come from remote observation using a technique known as spectroscopy, which relies on using instruments called spectroscopes which take incoming light and break it up into its component parts, producing an effect similar to a rainbow. By examining the resultant spectra from the Sun and indeed other stars we can make reliable deductions about conditions deep inside.

We can also learn from the spectrum what the Sun is made of, by looking at its absorption lines. These dark lines in the spectrum exist because of the presence of different gases. If you imagine looking at a pure white light source through a spectroscope then you would see a perfect spectrum with the colours running from red through to violet. If you now somehow managed to place a cloud of gas in between the light source and your spectroscope then dark absorption lines would appear. The lines exist because of the presence of the atoms of gas, or, more precisely, the presence of electrons around the nucleus of the atom. When the incoming photons of light strike the electrons in their orbit around the nucleus, they get absorbed, giving the electrons a bit of energy and boosting them to a higher energy state. We 'see' this process in the existence of the dark absorption lines, and the arrangement and positioning of the lines tells us which gases are present. From studying the Sun in this way, we know that it is made up of 91.2% hydrogen, 8.7% helium and tiny amounts of oxygen, carbon, nitrogen, silicon, magnesium, neon, iron and sulphur.

Knowing what the Sun is made of has enabled us to categorize it as a third-generation star, which astronomers confusingly call a Population I star. The first generation of stars formed early in the evolution of the Universe and lived fast lives. Because they were the first stars that formed, they were almost entirely composed of hydrogen and helium with tiny amounts of lithium too, as these were the only elements synthesized during the Big Bang. As the stars evolved, nuclear fusion deep in their core produced heavier elements through fusion, which are known as metals. After just a few million years the first-generation stars died, seeding the Universe with the heavier elements, and it was out of this material that the second generation of stars formed. They are distinguishable because they have some quantities of the heavier elements produced from the fusion inside the first-generation stars. The cycle continued, and fusion inside the second-generation stars produced even more heavy elements which were eventually liberated to go on to form the third-generation stars. Spectroscopy allows us to discover how many heavy elements are present inside stars and the proportions tell us whether they are second- or third-generation stars. The Sun, then, is a third-generation star because of the quantities of heavy elements that have been found inside it.

After the death of second-generation stars in our region of the Galaxy, the resultant clouds of old stellar material started slowly to coalesce again under the force of gravity. As the temperature and pressure inside the cloud increased beyond a critical level, nuclear fusion began to take place – the process whereby atoms are crashed together and in doing so form different elements. In the case of the process inside the Sun that we see today, for every four hydrogen atoms that fuse together, they produce one helium atom. There are other particles involved in the process such as electrons and neutrinos, but more importantly two of the by-products of fusion are light (photons) and heat, and it is this heat and light formed in the core of the Sun that warm and brighten our world.

You won't see any sign of this from your vantage point in the *Kaldi*, but before they can escape from the Sun, they have a perilous journey to freedom. Heat has the easier job as it simply finds its way to the photosphere of the Sun through radiation and then convection (hot material rises and cooler, less dense material sinks). Light, on the other hand, has a much harder job of it because of the density of solar material. Instead of travelling in a straight line, light weaves around like a drunk staggering home, making random turns left and right, bumping into things and looking for things to hold on to. Knowing how far a drunk has to travel, the length of his stride and the number of steps he takes, it is possible to estimate how long it would take for him to make the journey. We can apply the same concept to photons of light trying to get out of the Sun that bump into protons and electrons along the way. Plugging in some estimates we get anything from 400,000 to 1 million years for the average time it takes a photon of light to travel a distance of just 695,500 kilometres. That means the average photon's rate of progress inside the Sun is 0.06 millimetres per second instead of the usual 300,000 kilometres per second that we see in the vacuum of space. Once outside, though, it rockets along and covers the 150 million kilometres to Earth in just 8.3 minutes.

The core is the real powerhouse of the Sun as it is here that the vast majority of the nuclear fusion is taking place. It is about 347,750 kilometres in diameter with a temperature of 15.7 million degrees Celsius and a density about 150 times that of water. The energy generated in the core of the Sun through fusion can be compared to the amount of heat generated inside an active compost heap. This may seem a little surprising, given that the heat of an average garden compost heap is clearly nowhere near the 15 million degrees the Sun reaches. But the energy production per unit of volume in the Sun is the same as the energy production per unit of volume inside the compost heap. The incredible power of the Sun is due to its enormous size rather than the power production per unit of volume.

Surrounding the core is the radiative zone with a depth of about 139,100 kilometres. In this region the density of the solar material and its temperature is sufficient for heat to be transferred through radiative means. The same transfer mechanism allows us to feel the heat of the Sun from Earth despite being separated by 150 million kilometres of empty space. By the time the top of the radiative zone is reached, temperatures have plummeted from the searing 15 million degrees of the core to a rather more temperate 2 million degrees. Like the core of the Sun, the zone is thought to rotate in a fairly uniform fashion, unlike the next layer above which is known as the convective zone. It rotates in a differential way, which means the different regions rotate at different speeds. In order for there to be a transition between the two there is a layer that sits between them called, unsurprisingly, the transition layer. It is here that there are changes in rotation: successive layers glide past each other, converting the fluid motion of the convective zone at the top to the uniform rotation of the radiative zone below. There is a school of thought that suggests the movement in this layer generates a magnetic dynamo effect that is the cause of the Sun's magnetic field.

So, above the transition layer is the convective zone, and here the temperature and density is lower than in the radiative zone so conditions are better for the transfer of heat energy. The convective layer extends from the top of the transition layer at a depth of about 200,000 kilometres to the photosphere. Solar material at the bottom of the convective zone will pick up heat from the top of the transition layer and expand. As it expands it will decrease in density and rise. Once it gets to the photosphere, the rising material radiates its heat out into space and cools. As it cools it shrinks, which leads to an increase in density, so it sinks back down into the convective zone. This process builds convection cells by which heat is carried to the surface in much the same way as convection cells in a saucepan of water carry heat from the bottom of the pan to the top. The convection cells on the Sun can be seen when observed in visible light and are known as granulation, because they give the Sun a somewhat speckled, granular appearance.

All layers below the photosphere are opaque to light, and it is only when photons reach the photosphere that they are free to head out into space. The photosphere, whose name comes from the ancient Greek meaning 'sphere of light', is only a few hundred kilometres thick and has a temperature of about 6,000 degrees. There is a significant reduction in negative hydrogen ions (hydrogen atoms that have gained an extra electron) which easily absorb photons of light, whereas other layers of the Sun have them in abundance.

Granulation is not the only solar phenomenon that can be seen in visible light. Another effect that can be seen is the so-called limb darkening, where it seems to be brighter in the centre and gets progressively darker the nearer you get to the edge of the disc. There are two fundamental concepts which lead to this phenomenon: first, the density of the material decreases as distance from the centre increases, and second, the temperature also reduces with distance from the core. By looking at the centre of the Sun's disc we are looking straight into the Sun so our line of sight takes us deeper into the photosphere, whereas when looking at the limb, our line of sight is at more of a glancing angle so we are not looking so deeply. Given that the Sun is hotter the deeper you go and that hotter material is brighter, it's no surprise that by looking deeper you are looking at brighter regions of the Sun.

Granulation and limb darkening are subtle effects when viewed from Earth, though from your vantage point on the *Kaldi* they are much more prominent. More obvious are the aptly named sunspots that grace the surface of our nearest star. They vary in size from just a few tens of kilometres across to up to 150,000 kilometres – large enough to dwarf the Earth and to be seen with the naked eye (with appropriate protection). While they appear black they are actually still very bright features: if you could somehow separate them from the bright disc of the Sun they would still shine brighter than the full Moon, even at that enormous distance.

Like many features on the Sun, their appearance is ultimately tied up with the nature of plasma and magnetism. We have already seen that plasma can be moved by magnetic fields, and that this process slowly accelerates your spaceship as plasma is channelled by magnetic fields and expelled out of the VASIMR engine. We've just looked at the concept of differential rotation, and in the case of the Sun, plasma around the equator rotates faster than plasma around the polar regions. As the plasma rotates, it drags the magnetic field around with it, increasing the stresses it experiences. When the levels of stress get too high, the lines buckle and occasionally burst through the surface, inhibiting the convective activity leading to a cooler surface temperature. Their temperature, usually around 2,000 degrees cooler than the surrounding photosphere, means that the points where the field lines exit and re-enter the photosphere appear darker than the surrounding solar material.

Under observation the detail of larger sunspots can be amazing and very much resembles the appearance of the experiment you may remember from your school days with iron filings and bar magnets. There is a central darker region known as the umbra and it is here that the magnetic field lines are almost vertical with respect to the photosphere. Surrounding the umbra is a lighter region known as the penumbra where the magnetic field lines have a more inclined angle. Generally sunspots come in pairs, with one spot marking the point where the field bursts through the photosphere and the other the point where it dives back in again. If you study the magnetic properties of sunspots in a pair then, just like a bar magnet, there will always be one that seems to represent the north pole and the other the south. Interestingly, if you then compare a sunspot pair with all the others in the same hemisphere of the Sun, they will all have the leading sunspot representing the north or south pole with the opposite trailing. The opposite is then true for the sunspot pairs in the other hemisphere, so they are clearly linked to the global magnetic field. The quantity of sunspots seems to vary too, from sometimes none through to ten or more.

The variation in the quantity of sunspots visible on the Sun is known as the solar cycle, with the most sunspots being visible at solar maximum and the least at solar minimum. At the start of a new cycle, the sunspots tend to be restricted to higher latitudes of the Sun, around the polar regions, but as the cycle progresses they tend to migrate further towards the equator as their numbers increase. The cycle has an eleven-year period from maximum to maximum which ties in with the state of the magnetic field: when the field is at its most stressed we experience maximum sunspot activity and when it has eventually snapped back to a state of quiescence we experience solar minimum. This cycle is predictable but irregular with the quantities of sunspots visible at maximum changing. For example, between 1900 and 1960 the trend was generally an increase, but in the years that followed there seemed to be a gradual decline. Periods of particularly low sunspot activity have also been observed historically, such as the Maunder Minimum between 1645 and 1715 when sunspots became unexpectedly rare. The cause of this variation is still not well understood.

Above the photosphere is the atmosphere of the Sun. There are four different regions in this atmosphere. The lowest area, found just a few hundred kilometres above the photosphere, is known as the chromosphere, and above this is the transition region. This is then surrounded by the corona, and finally there is the upper region known as the heliosphere which extends out beyond the orbit of the minor planet Pluto. The coolest region of the Sun lies just between the photosphere and chromosphere and is known as the 'temperature minimum region'. The temperature in this thin region is around 4,000 degrees and it is here that water molecules and other simple compounds are found. These can be detected through spectral analysis using a spectroscope, as explained earlier.

Above the temperature minimum region is the chromosphere, which literally means 'sphere of colour'. One of its key characteristics is its low density, typically about 0.0001 times the density of the photosphere. The density drops with increasing distance from the Sun, but interestingly the temperature profile changes with a slow reduction from the inner boundary before it starts to rise back up towards the outer edge with temperatures in the region of 35,000 degrees. In contrast to the spectrum of the photosphere, which displays absorption lines, the spectrum of the chromosphere displays emission lines. We have already seen how absorption lines are the result of electrons in orbit around the nucleus of an atom absorbing specific wavelengths of light as they jump to a higher energy state. Their counterpart, the emission lines, are seen when the electrons hop back to their usual energy level and emit light in certain wavelengths, and it is this type of spectrum that is seen when the chromosphere of the Sun is observed. Instead of looking like the familiar rainbow with dark lines superimposed, the spectrum of the chromosphere is just a series of colourful lines. The most prominent lines in the solar chromospheric emission spectra are from hydrogen-alpha atoms with a wavelength of

656.3 nanometres (there are 1 million nanometres in a millimetre). Light at that wavelength is in the red part of the spectrum so the chromosphere has a distinctly red colour to it, although it is not easy to see unless during a solar eclipse.

The chromosphere is home to some of the most striking and interesting features on the Sun, but studying them from Earth requires either the photosphere to be blocked from view by the Moon during an eclipse or a specially filtered telescope that can study the hydrogen-alpha emissions. Spicules are by far the more common features – long fingerlike tendrils radiating out from the Sun. They rise to the top of the chromosphere before dying back down again after about ten or fifteen minutes. A little more spectacular but less common are the solar prominences and filaments. Prominences can be seen around the edge of the Sun where giant arcs of gas have been dragged off the photosphere by magnetic disturbances, usually above sunspots. Filaments are the same event but when viewed from above, so they look like dark lines staggering across the photosphere. Just like spicules, they reach high into the upper reaches of the chromosphere, at altitudes of up to 150,000 kilometres. Even at your close range they are difficult to spot due to the extreme contrast from the brightness of the photosphere, although spicules and prominences can be seen when superimposed against the inky blackness of space.

Surrounding the chromosphere is the thin layer (just 200 kilometres thick) known as the transition region. Here the temperature matches that of the upper regions of the chromosphere at around 35,000 degrees but rises rapidly to about 1 million degrees. It is only possible to see the transition region using space telescopes sensitive to ultraviolet radiation which is largely blocked by the Earth's atmosphere. Through observations from space and spectral studies it is possible to see that the helium in the layer is completely ionized. The fully ionized helium means that the helium atoms have lost all their electrons, and in this state the radiative cooling process is severely limited so the region is unable to cool down, leading to very high temperatures. Helium in its partially ionized state is found in the upper regions of the chromosphere but with a little extra heat it becomes fully ionized, and it is this process that is critical for the formation of the next layer of the Sun's atmosphere, the corona.

The corona is without doubt the most spectacular part of the Sun's atmosphere and is one of the most interesting regions to study. It extends out into space for many millions of kilometres but is only observable from Earth during a total solar eclipse or when the Sun is looked at through a coronograph, which simulates an eclipse. Light that can be seen coming from the corona actually comes from three different sources. The first is the result of light being emitted from ions in the coronal plasma. As free electrons reattach to atomic nuclei, they emit light which can be seen as an emission spectrum – hence the name the 'E-Coronal emission'. Then there is the F-Corona, which is named after the Fraunhofer absorption lines which are made visible due to light from the photosphere bouncing off dust particles (Joseph von Fraunhofer was the nineteenth-century German optician who invented the spectroscope and discovered the lines). The F-Corona is the light often referred to as the rare zodiacal light that can be observed as a diffuse glow in the sky, usually seen on the horizon just after sunset or just before sunrise, but from very dark moonless sites it can be seen stretching around the entire ecliptic. The K-Corona is the final source of coronal light and is the result of photospheric light being scattered off electrons that are not attached to an atomic nucleus. The absorption lines that are seen in light from the photosphere are absent in spectra from the corona as a result of the motion of the electrons scattering the light, causing the lines to become wider so that they become invisible. This phenomenon can be used to estimate the temperature of the corona because the broadening – known as 'thermal Doppler broadening' – is dependent on the wavelength of light, the mass of the emitting particles and their thermal motion. The 'K' in K-Corona comes from the German *Kontinuierlich*, which means 'continuous', in reference to the corona's continuous spectrum without absorption lines.

One of the biggest challenges facing solar physicists is understanding and explaining the extremely high temperatures found in the solar corona. We know that the transition region below reaches temperatures in excess of 1 million degrees and this is easily explained by the ionized helium, as we have just seen, but temperatures in the corona have been found to reach a staggering 3 million degrees. One possible solution lies in an induction process from the magnetic field of the Sun where the relative motion of the plasma and the ever-changing magnetic field causes the heating. But for now, the exact cause is not understood.

One thing that is clear is that the corona, like many areas of the Sun, is constantly changing. We have already looked at the solar cycle and how this varies over an eleven-year period, and the corona seems to follow this pattern too. During the period when the Sun is most active the corona seems to cover the majority of the Sun, but during quiet times it seems only to be present around the equatorial regions, which would suggest it is being manipulated by the magnetic field. A great example of how it is affected in this way is the beautiful coronal loops which are a direct result of the disturbed and ever-changing magnetic field. We have seen how sunspots are the result of the magnetic field bursting through the Sun's visible surface; coronal loops are the manifestation of this process in the corona, where the loops trace out the path of the magnetic field looping through the coronal plasma. The density of the plasma in the corona is about 10 billion times less than the density of material in the photospheric plasma so features like coronal loops are considerably fainter.

One feature which seems to permeate through all layers of the Sun's atmosphere is the solar flare, which can often be seen as a sudden brightening on the solar disc or around the limb. A significant amount of study has gone into these flares which represent a sudden burst of energy lasting just a second or two but releasing about 30 million times more energy than all the nuclear missiles ever detonated. They seem to be related to changes within the magnetic field, for when magnetic field lines interact with plasma the lines seem to become grouped or bundled together into domains which connect from one place to another. Even though the domains will experience high levels of distortion from movement of the solar plasma, they still tend to retain their unique geometrical properties when compared to other magnetic domains nearby. Different bundles are separated from each other by curved boundaries known as separatrices which border regions of plasma that behave in different ways. It seems that magnetic field lines from different bundles that were not previously connected can break through the separatrices and splice together in a process known as 'magnetic reconnection'. In doing so, they convert magnetic energy that has been stored up in the magnetic field for days into kinetic and thermal energy in just a matter of minutes. The reconnection process heats the plasma to several millions of degrees and accelerates electrons and ions to speeds near the speed of light, and the whole process releases a burst of energy and light which we see as a flare.

These flares tend to erupt in regions around sunspots where magnetic fields have burst through the photosphere and extend up into the corona. It is thought that the same process is responsible for the even more violent coronal mass ejections that are occasionally observed which send an extra burst of the solar wind out into the depths of the Solar System. We have not yet identified a direct link between flares and these ejections but they do still seem to come from the same cause, magnetic reconnection. Coronal mass ejections differ from flares in the huge amount of material that gets thrown out from the Sun into the depths of space and in different observed properties, usually a bright leading edge. The ejected matter can reach speeds of up to 750 kilometres per second, although there is a slower component whose speed is a mere 400 kilometres per second. Both the slow and the fast solar wind contain electrically charged particles but they are at different temperatures and come from different parts of the Sun. The slow solar wind seems to originate from the equatorial regions while the fast wind comes from coronal holes which literally are holes in the Sun's corona around its polar regions. As we have seen, it is the slow solar wind which is responsible for the beautiful auroral displays we see on Earth in far northern and southern latitudes. So the effects of solar wind are far-reaching. While coronal mass ejections produce a fresh burst of solar wind, the Sun is actually giving off a near constant flow, even during quiet periods.

The corona represents the outermost layer of the atmosphere of the Sun, but in reality it extends very much further. Surrounding it is the much more rarefied heliosphere, which reaches to the edge of the Solar System. Indeed the Solar System is defined by the heliosphere. The point at which the influence of the Sun is indistinguishable from the influence of neighbouring stars is known as the heliopause. The heliosphere can be thought of as a giant bubble filled with solar material, primarily from the solar wind and the extended solar magnetic field. It forms a bubble because of the outward flow of pressure from the solar wind which pushes against the inward pressure from interstellar space, and the point at which the two balance creates the boundary and the heliopause. The exact location of the heliopause has been a matter of great debate for decades, but in 2012 Voyager 1 finally discovered its location around 18 billion kilometres from the Sun, which is about 123 times the distance between the Earth and the Sun. The exact shape and location of the heliopause does change though with the fluctuations in the solar wind and pressures caused by interstellar gas as the Sun moves around the Galaxy.

It is fair to say that we have a pretty good understanding of the Sun, but there are still many mysteries left to be solved, not least the extended periods of relative dormancy like the Maunder Minimum or the cause of the incredible temperatures in the solar corona. What we do have a good grasp of, however, is the likely fate of the Sun. For the majority of its life it experiences two forces: an outward pushing force generated by the fusion process in its core, known as 'thermonuclear pressure', and an inward pulling force from the force of gravity. For the last 4.5 billion years the Sun has sustained a balance between the thermonuclear force and the force of gravity, but this will not always be the case. There is actually a tiny imbalance between the two which is causing the Sun to expand very slightly: it has grown in size by 6% over the last 4.5 billion years. There is enough hydrogen in the core of the Sun for it to continue its current rate of fusion for another 5 billion years or so, which means it is about halfway through the stable phase of its life. It is thought the Sun will continue to expand gradually over those 5 billion years, but after that things will change.

Eventually the core of the Sun will run out of hydrogen, the thermonuclear force will momentarily decrease, and the core will start to collapse as the force of gravity takes hold. While this is happening there will still be some hydrogen fusion taking place in the upper layers of the Sun, but as the core contracts it will heat up, causing the temperature in the upper layers to increase too. As the temperature increases the outer layers will expand, causing the Sun to grow in size more quickly: it will double in size in just 500 million years and turn into what is termed a sub-giant. After that the expansion will continue, but at an even faster pace. Over the next 500 million years it may swell to 200 times its current size and be over 1,000 times more luminous. The Sun will become a red giant. Exactly how big it will get at this stage in its evolution is uncertain but its surface will almost certainly extend beyond Mercury and Venus, and there is a chance it may also swallow up the Earth.

The compression of the core will not only increase its temperature, it will also increase the pressure so that eventually conditions will be right for helium fusion to start, leading to the production of carbon in the core. The layers of different gases inside the Sun will now resemble an onion with several distinct layers: hydrogen outer layers which surround a helium shell around a growing core of carbon. It will spend about a billion years of its life like this before the core violently ignites in an event known as a helium flash, and then the Sun will shrink to about ten times its current size. In this almost catastrophic event it will be the quantum forces in the core that will resist the crushing force of gravity rather than thermonuclear pressure. These forces exist between the particles in the atom and are actually incredibly strong – strong enough to hold back the force of gravity. Because thermonuclear pressure alone is not enough to expand the material and thus cool it, the fusion process will go on unchecked and a nuclear chain reaction will explode through the core. In a brief few seconds, the core will put out as much energy as the entire stellar output of an average spiral galaxy, but this will be absorbed inside the Sun. The temperature will eventually grow so high that thermonuclear pressure will cause the Sun to expand once more, cooling the material and checking the runaway nuclear reactions in the core. It will stay like this for about another million years, fusing helium into carbon deep in the core.

When the helium in the core is exhausted the Sun will once again expand, just as it did when the hydrogen fuel ran out. The expansion and increase in luminosity will happen much faster this time and after about 20 million years it will become increasingly unstable. Every 100,000 years or so the Sun will pulse, each successive one becoming more and more extreme with more and more material escaping into space. Estimates suggest that just four or five pulses will be enough for the Sun to lose its outer layers to space and enter into its planetary nebula phase. This will take about half a million years to form and will mean the rest of the Sun that has been left behind will have about half the mass it has today. The remainder will form an expanding shell of gas which any external observer might see like the Ring Nebula in the constellation of Lyra. Eventually, after about another 10,000 years, the expanding shell of material will dissipate into space and go on to form the next generation of stars and planets. But even that is not quite the end of the story. After the core of the Sun has been exposed with the formation of the planetary nebula it will then cool to become a white dwarf.

Eventually it will cool even further and stop emitting light, at which point it will fade away and become a large black stellar corpse.

So, fortunately for us, the death of our Sun is many billions of years away. None the less, for Earth the clock is ticking and our time is limited. The Sun's slow but steady increase in luminosity and temperature means that within a billion years our atmosphere will warm and temperatures will be high enough for high-altitude water not to freeze, allowing water molecules to reach the very top of the atmosphere and escape into space. In reality, our time may not be solely limited by the Sun's extinction process but by other celestial events too. Perhaps the exotic gamma ray bursters whose intense radiation permeates the Universe, destroying any planetary atmosphere in their wake, will put an end to life on Earth, or maybe a wayward asteroid will deliver the final blow. One way or another we will need to find a way of getting off this planet before that time comes, and that is why journeys of exploration like the *Kaldi*'s are so important. For now, though, our local star is in the prime of its life.

For your close fly-by there are a few things that really should be considered. The first is that a solar fly-by will grant you no net speed increase. As we have seen, the extra boost a spacecraft gains from a planetary fly-by is its speed relative to the Sun. As a spacecraft approaches a planet it speeds up, and after the encounter it slows down without any net speed gain relative to the planet. Relative to the Sun, however, the spacecraft steals some of the orbital movement of the planet, slowing the planet a tiny amount and speeding up the spacecraft. Performing a close pass of the Sun provides no increase in speed, though it will help to set you on your path to Mercury.

We have already seen, too, how the planets move around the Sun, and this allows you to accelerate your spacecraft, but the Sun moves as well. The Sun and the entire Solar System is in constant movement around the centre of our Galaxy and it is this movement that can be traded for velocity if we intend to travel among the stars. The speed of the Solar System around the Galaxy is an incredible 828,000 kilometres per hour and at that speed it will take about 230 million years to complete one orbit. It is a sobering thought that since our first recognizable ancestors walked the Earth, we have completed only one thousandth of an orbit, which does rather put human life in perspective.

The Sun sits at a distance of about 100,000 light years from the centre of our Galaxy which is thought to be home to a supermassive black hole. These fascinating objects have never been observed directly but we can infer their existence by observing the effects they have on the surrounding regions of space. Unlike our Sun, larger stars die much more dramatically when they literally blow themselves to pieces at the ends of their lives. The remaining stellar corpse may be a super-dense neutron star, pulsar or the even more exotic black hole.

When observed from intergalactic space, our Galaxy looks like two fried eggs stuck back to back with the yolks at the centre representing the bulging core and the egg whites representing the plane within which we find the galactic spiral arms. Some observations have hinted that we do not live inside a spiral galaxy but instead a barred spiral, where a strange bar-like structure seems to cross the nucleus. The winding spiral arms seem to emanate from the loosely defined bar of the Galaxy, although there is no satisfactory explanation for why they exist. We can certainly tell that spiral galaxies and barred spiral galaxies like our own seem to be alive with star formation, suggesting that they are young, whereas the older stars and lack of stellar birth in the elliptical galaxies hint at some kind of evolutionary process that all galaxies follow.

The Sun's position in our Galaxy is near one of the four spiral arms known as the Orion Arm, which is found between the Perseus and Sagittarius Arms. Despite their appearance, the arms in spiral and barred spiral galaxies do not actually rotate; instead the Galaxy rotates through them. They are thought to be some kind of density wave phenomenon caused by the movement of stars and gas in just the same way that a traffic jam forms on a free-flowing motorway simply because of the movement of the cars. Over time, we will slowly move away from the Orion Arm which our Solar System is currently close to and head towards another spiral arm. Compared to other barred spiral galaxies, our own is pretty average in shape and size. When measured from one side to the other it spans a staggering 100,000 light years and is believed to be about 28,000 light years from the galactic centre.

At the start of the chapter we looked at the technology that would give us a view of the Sun and afford us some protection against its rays. Unfortunately the solution is not yet a complete one, because as you get closer to the Sun and temperatures increase the gold film on your windows is at risk of melting. It has a melting point of just over 1,000 degrees, and we know that the temperatures in the corona are well in excess of 1 million degrees, sometimes reaching three times that. Even the coolest region of the Sun's lower atmosphere is way above the gold film's melting point, so you can't fly too close. The outer layers of the corona are cooler so it is at least theoretically possible to skip through there, and your journey will take you to about 2 million kilometres from the photosphere, which is about twenty-eight times closer to the Sun than Mercury. At that distance the temperatures could be well in excess of 1,500 degrees. It is difficult to accurately predict the conditions but we must come up with a solution for keeping the spacecraft cool enough for you to survive.

The ideal material for such an extreme heat shield is carbon, which has the highest melting point of all the elements at 3,500 degrees. That should make it capable of withstanding anything the outer corona can throw at you. The space shuttle very successfully utilized carbon tiles on its underside to protect it from the high temperatures of re-entry into the Earth's atmosphere. By employing a carbon heat shield in the same way it will be possible to absorb the heat without damage, and with temperatures on the rear of the *Kaldi* reaching only a few hundred degrees, the heat can dissipate off into space, keeping the craft (and you) cool.

Solar heating can be a problem even for spacecraft in orbit around the Earth because the sunlit side can still reach a few hundred degrees while the area in shadow will be very cold, often too cold for instruments to be able to operate, so one of the challenges facing space engineers is working with these extremes of temperature and harnessing them to keep instruments functioning. One way this can be achieved in colder environments is by adapting ecologically friendly solutions on Earth such as solar water heating systems. Water-filled pipes can be run around the spacecraft and the Sun's heat will warm the water which can then be stored and released when required to provide a sort of central heating system.

Radiation is one of the other major problems facing your journey around the Sun. Astronauts on board the International Space Station are protected to some extent by the Earth's magnetic field, but even so they receive a higher dose. An annual dose for an astronaut on board the station is about 150 millisieverts per year, whereas their family and friends back on planet Earth receive an average dosage of just 0.3 millisieverts per year. (A millisievert is a term you may not have come across before. It refers to ionizing radiation doses and it measures the biological effect of the dose. In particular it refers to the quantity of radiation experienced over a given time period that has sufficient energy to cause an electron to escape from the bond of an atom and therefore ionize it.) Long-term exposure needs to be kept to a minimum, but the body can cope with short-term doses: the astronauts on board the Apollo missions to the Moon, for example, experienced around 1.2 millisieverts per day, much higher than the ISS astronauts, but for a few days it was an acceptable level. Clearly the length of your trip around the Solar System means you are at risk from high levels of radiation, and certainly near the Sun that will increase significantly. And it is not just high-energy particles from the Sun that you need to worry about on your journey: you must also keep an eye out for cosmic rays from deep space.

By spending a year unprotected in space the life expectancy of a traveller would reduce by a couple of months, so some form of protection from radiation is definitely needed. We looked earlier at using superconducting magnets, and this seems to be the most efficient solution, certainly compared to lining the spacecraft with thick lead which would add weight and bulk. As you approach the Sun, not only do heat and radiation increase but so does exposure to ultraviolet wavelengths. We are all familiar with this from our beach holidays: prolonged exposure to sunlight leads to a browning or reddening of the skin which we know as either a suntan or sunburn. Protection from ultraviolet radiation is fairly easy to resolve, though, with reflective surfaces over the entire craft. So, with a carbon shield to cope with the heat, a reflective surface to deal with ultraviolet radiation and a magnetic field to block out as much radiation as possible, you are as protected as you can be.

One final danger that does need consideration as you swing past the Sun is the extreme tidal force exerted on the ship. To understand the dangers we only have to look at the ill-fated Comet ISON (officially known as Comet C/2012 S1), which swooped past the Sun during the last couple of months of 2012 at a distance of just 1.8 million kilometres. The Sun's immense gravitational pull managed to overcome the gravitational force holding the comet together and it disintegrated. This is often the case for so-called 'sun-grazing comets'; their chances of survival chiefly depend on how well they are held together. There is a region around all objects that is known as the Roche limit, and at this distance the gravitational pull of the central body is likely to overcome the force of gravity holding an object together, leading to its destruction. The distance of the Roche limit also depends on composition, and for a comet in orbit around the Sun the limit is around 1.2 million kilometres. Clearly Comet ISON was not made of strong stuff. Your closest approach to the Sun is going to be 2 million kilometres, so while that is cutting it a bit fine, hopefully the strength of the *Kaldi* will be able to withstand the tidal forces exerted upon it.

Cooped up behind the protective layers of the spacecraft you can safely observe the awesome power and activity of the Sun, but what are the chances of taking a spacewalk from here? It would be amazing to be able to see the Sun right there in front of you, unhindered by the spacecraft's protective shielding. But if you were to venture out of the craft your chance of survival is pretty limited. The current space suit design will withstand temperatures of about 250 degrees, which enables humans to carry out spacewalks up to a distance of about 5 million kilometres from the Sun. Any closer and the temperature inside the suit would rise beyond tolerable levels until dehydration would set in and the occupant would pass out before eventually dying from heatstroke. Use the RSU if you must, but you've had a pretty spectacular look at the Sun already, and it is a monstrously fierce thing, so best to stay inside the sanctuary of your ship and get on with the next leg of the journey.

FOUR

The In-Hospitality Suite

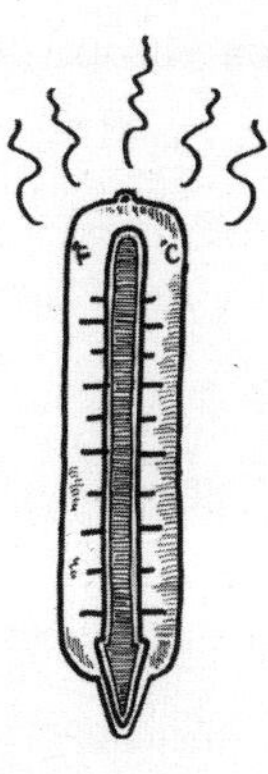

WHILE THE SUN IS without doubt the most inhospitable place in the Solar System, the inner planets Mercury and Venus are certainly not places you would want to head to for your summer holidays. Travelling around the Solar System on a journey like yours is a huge undertaking, but fortunately the gravity induced by the rotation of your living quarters means that you can at least enjoy a few home comforts.

Baths, for example, can be enjoyed in the normal way because water will act in a way you are used to. Where astronauts are living in zero gravity environments, such as on the International Space Station, personal hygiene tasks are less than easy because water does not flow. If you were to take a sponge that has been immersed in water and squeeze it in zero gravity the water would squeeze out of the sponge but just cling to the outside until you stopped squeezing, at which point it would be reabsorbed. Similarly, if you were to pour water out of a container it would just form into spherical blobs that would float around. Despite these limitations it has always been essential to take personal hygiene in space seriously because the closed and cramped living conditions encourage germs and bugs to spread. To take showers in space, astronauts in zero G have to climb into a cylinder that is effectively a big plastic sleeve which stops the water from floating away. Water is then sprayed on to them from a nozzle, and once finished, a vacuum is used to suck up the water from their skin. Early shuttle astronauts had even less luxury as showers were never installed; instead they used wet wipes to clean their bodies. Because washing facilities were non-existent, their clothes were disposable and would be thrown away after about a week's worth of wear. Thankfully baths and showers are easy enough to enjoy as you head along on your journey.

Owing to their close proximity to the Sun, Mercury and Venus are drenched in heat and intense solar radiation, making conditions pretty grim. Mercury is the nearest of the planets to the Sun and because of this it is only ever visible in the sky from Earth in the morning or evening twilight. Venus is the second planet out, but, for reasons we will look at later, it is hotter than Mercury. According to Kepler's laws of planetary motion, the closer a planet is to the Sun the faster it will travel, and in the case of Mercury it rattles around completing one orbit of the Sun in just 88 Earth days, while Venus takes a little over 224 days.

Mercury and Venus are often referred to as the inferior planets since they orbit the Sun closer than Earth, but they are not the only inner planets, which comprise Earth and Mars too. The four planets share similar properties: they are all rocky bodies composed mainly of silicates in the crust and mantle, and iron and nickel in the cores, and their features differ widely from those of the outer gas giant planets. For starters, they lack significant quantities of moons. Mars has two, Earth has one and Mercury and Venus have none, compared to the outer planets, each of which has anything from just over ten to in excess of fifty. The inner planets also lack a ring system, and one of the most fundamental differences is the extent of the atmosphere that surrounds them. The outer planets are almost entirely made of gas, hence their collective name, their gas atmospheres contributing well over 90% of their composition. To understand the nature and cause of this difference we need to look back to the origins of the Solar System. When it formed over 4.5 billion years ago, it grew out of a vast cloud primarily made of hydrogen, but there were other elements present. As we saw in the previous chapter, the Sun formed out of the collapsing cloud and started to kick out incredible amounts of energy which pushed against the surrounding gas cloud, forcing the lighter chemicals to the outer reaches of the Solar System. It was out of this much lighter gas that the gas giants grew while the heavier elements, which could resist the pressure from the Sun, remained in the inner Solar System and formed the rocky planets Mercury, Venus, Earth and Mars.

Getting to Mercury and Venus will be relatively simple as unlike the outer planets you will not be fighting against the immense pull of gravity from the Sun trying to hold you back. One of the first spacecraft to travel to the two innermost planets was Mariner 10, which was launched in November 1973. Just three months later, in February 1974, it passed Venus at a distance of 5,768 kilometres, and Mercury a month later at a distance of just 703 kilometres. On its way to the two hostile worlds Mariner had to make a number of mid-course corrections, and to ensure the right ones were made it used the stars to navigate.

Bright stars like Canopus are often used by unmanned spacecraft for navigation purposes. This is usually done with a device known as a star tracker which uses any number of stars from fifty-seven or so of the brightest candidates. Not only do they help with identifying location, they are of equal importance in helping to determine orientation. You may well know your position in the Solar System but it is of the utmost importance to be facing in the right direction before making course corrections, and without being able to see your destination this can be difficult. Newton's laws of motion are key here, in particular the third law, which states that every action has an equal and opposite reaction. Imagine you are facing forward, in the direction of travel, and you fire the thruster on the left or port side of the craft; the path of the spacecraft will adjust slightly to the right. But if the spacecraft is facing backwards, looking in the direction it has just come, then firing the left thruster will result in its path being adjusted to the left. There are an almost endless number of variables when it comes to making course corrections; getting them right means understanding exactly where the spacecraft is, understanding exactly in which direction it is facing, and deciding exactly how much thrust is required to make the correction.

The journey to Mercury, the first of the inner planets you're going to visit, takes just a couple of months. There was a time when Mercury was the second smallest planet in the Solar System, beaten only by Pluto whose diameter is about 2,519 kilometres smaller. Things changed in 2006 when the International Astronomical Union finally decided to define what a planet actually was. Until then there was no definition. Without any hard and fast rules that objects had to adhere to in order to be considered a planet, our Solar System had kept on growing, from nine planets, to ten, then eleven with the discovery of rocky bodies like Quaoar and Sedna. The definition of a planet was finally agreed and passed at a conference of the IAU on 24 August 2006 and it meant that poor Pluto got demoted to the status of a mere minor planet.

There are three criteria an object must satisfy in order to be classed as a planet, and the first is that it must be a celestial body in orbit around the Sun. It must also have sufficient mass for gravity to assume hydrostatic equilibrium – in other words, it must be nearly spherical in shape. Both Mercury and Pluto meet these requirements, but it is the third criterion that strikes a blow for the status of the latter: it must be gravitationally dominant and have cleared the neighbourhood around its orbit of objects of comparable size. Pluto has not cleared its orbit and become dominant as it is accompanied by other rocky bodies. We will come back to all of this in more detail when you are in the depths of the Solar System, but for now, and until any other discoveries change the landscape of the Solar System, Mercury is the smallest of all the major planets.

As well as being the smallest planet in the Solar System, Mercury is also the fastest: as we have seen, it completes one full orbit of the Sun in just 88 Earth days. As it travels around the Sun, it rotates on its axis very slowly and in a way that is unique in the Solar System. When viewed from a distant star, it rotates only three times for every two orbits it completes, but when viewed from the Sun which rotates as Mercury orbits, it appears to make one rotation on its axis for every two orbits. That means that if you lived on Mercury you would see a sunrise and a sunset once every two years.

There is an even more bizarre effect that can be observed from some parts of Mercury: the Sun can be seen to rise, slow down, stop, and then head back in the direction it just came from before setting over the horizon it just rose from. This happens just before Mercury's closest approach to the Sun when its angular orbital velocity matches and then exceeds its angular rotational velocity and, a few days after closest approach, the Sun resumes its normal motion. The Sun does generally move slowly across the sky when this happens, but who could resist the chance to pop down to the surface to see it for oneself? The place to go would be the point on the equator where the Sun is overhead and at its nearest point, the perihelion. From there the spectacle could be enjoyed, but it would take about sixteen Earth days to watch the Sun slowly pass overhead, then stop, move back in the other direction and pass overhead again before stopping for a second time and resuming its usual motion, and passing overhead for a third time.

Something that will become really obvious to you once you've approached Venus for a good look is how different Mercury appears. Venus is enveloped entirely by a thick dense cloud that restricts the view of the surface. Mercury, on the other hand, seems to have no atmosphere at all so its surface can be seen in all its glorious detail. There is actually a very tenuous atmosphere known as an exosphere that loosely clings to the planet, but conditions on Mercury are not conducive to much more than that. The pull of gravity is not strong enough, given the high temperature and the constant onslaught of pressure from the Sun, for any significant atmosphere to evolve. The thin atmosphere that is there is made up of hydrogen, helium, oxygen, silicon and a number of other chemicals.

The conditions mean that atoms of gas can easily escape into space but they are constantly replenished from a variety of different sources. The hydrogen and helium atoms are thought to originate from the solar wind which has been trapped by the magnetic field, although the radioactive decay of atoms in the surface is also responsible for some of the helium. There is a wonderful process known as 'sputtering' where incoming energetic particles, usually ions or micrometeoroids, strike the surface and cause it to eject atoms, typically oxygen atoms. These then exist in the exosphere either as individual oxygen atoms or when combined with hydrogen atoms as water vapour. We know quite a lot about the Mercurian exosphere thanks to the Messenger spacecraft which arrived in orbit during 2011. Since that time it has carefully analysed the planet's chemical composition and studied the surface geology, the magnetic field and even probed the secrets of the core.

The lack of any decent atmosphere has led to Mercury experiencing the most extreme temperatures of any of the planets in the Solar System. On the Sun-facing hemisphere, the temperatures soar to 427 degrees, but on the night-time face they plummet to minus 173 degrees. On Venus, the dense atmosphere helps to equalize the temperatures across the globe, but on Mercury, heat quickly dissipates out into space. The only areas on Mercury to experience a fairly constant temperature are the polar regions which tend to stay at around minus 90 degrees due largely to the small tilt of the planet. The tilt of Mercury and indeed all planets is measured with respect to the plane of the Solar System known as the ecliptic, and it is along this plane that the planets orbit. The Earth, you will recall, is tilted over by just a little more than 23 degrees, which as we saw is what causes the seasons. The tilt of Mercury is a third of a degree, making it to all intents and purposes upright in its orbit. For that reason, the poles are largely unchanging in their presentation to the Sun, leading to a constant temperature. The lack of axial tilt also means that Mercury does not experience seasonal changes as many of the planets do. What it does have, however, is the most eccentric orbit of all the planets, so that at its nearest to the Sun, at perihelion, it is 46 million kilometres away, and at its most distant (aphelion) it is 69.8 million kilometres away.

Even though there are extreme changes in heat from day to night it looks very likely that ice exists on Mercury. There are craters in the polar regions that never receive sunlight at their bases and temperatures here remain well below zero. Radar has been used to bounce signals off these craters which reveal something highly reflective, which is very likely to be ice. It seems to be a familiar story that water is present in its solid state around the Solar System as it has been detected on the Moon, on Mercury, on Mars and on a couple of satellites of the gas giant planets. This makes longer-term human exploration and even habitation possible, at least in theory, because not only could water locked up in ice provide sustenance, but as we saw earlier the hydrogen and helium could be extracted to create fuel.

As you will notice, Mercury has an appearance very similar to that of the Moon. Thousands of craters are clearly visible, although vast lava plains seem much less obvious on Mercury. The lack of any appreciable atmosphere has allowed the planet's surface to be pummelled by countless impacts from tiny pieces of space rock. The craters within which the ice has been detected were created in exactly the same way as the craters we see on the Moon, by meteoric impact. Careful studies of the surface of Mercury have shown that there were two periods of heavy bombardment in its history, one shortly after its formation just over 4.5 billion years ago and another which seems to have come to an end about 3.9 billion years ago. For a planet that has had a history of meteoric impacts it is no surprise that craters are present in many different states, from those that appear quite fresh and new to others that are clearly very old and have been subsequently deformed by further impacts. The ejecta that is thrown out from the impact site does not cover as large an area as you saw around some craters on the Moon, chiefly because Mercury has a stronger gravitational pull which restricts the distance ejecta can travel.

The largest crater on Mercury is known as the Caloris Basin and measures 1,550 kilometres across, making it one of the largest impact craters in the Solar System. The impact of the 100-kilometre-diameter object that created it released so much energy that a ring of 2-kilometre-high mountains was forced up out of the crust to create the crater's edge. Studies of the floor of the crater reveal that it, like large craters on the Moon, is composed of solidified lava, which indicates that the impact likely cracked the crust to allow molten lava from below the surface to seep up and fill the basin. The relative absence of smaller impact craters across the floor suggests that this is a geologically young feature, perhaps having been created towards the end of the heavy bombardment era around 3.9 billion years ago. Spectroscopic studies have revealed that there seems to be a high concentration of sodium in the region around the Caloris Basin, suggesting that some of the cracks and fissures in the basin's floor may be facilitating the escape of gas from inside the planet which is then added to the tenuous atmosphere.

The naming of surface features on Mercury is a somewhat complex process. By decree of the International Astronomical Union, features including craters must be named after people who have made a substantial contribution in their field. New craters must be named after an artist who has been famous for more than fifty years and deceased for more than three years. Ridge features that stretch across the planet are named after scientists who have dedicated time to studying Mercury, and surface depressions commemorate high-achieving architects. Mountains get their name from various words for 'hot', for example Caloris Montes (Latin for 'mountains of heat'), while steep slopes or cliffs that occur from geographical faults are named after ships that have been involved in scientific exploration.

Aside from the numerous craters that pepper the surface of Mercury, you will see vast plains stretching across the landscape. The lack of significant cratering on them tells us that they are some of the youngest surfaces on the planet, but their origin is still unclear. They may be the result of large impact craters but equally could be volcanic in origin. The Caloris Basin is the only great plain that seems to bear definite signs of an impact origin, such as the fractures and ridges on the basin floor. The plains of Mercury generally resemble the lunar maria although they are much less prominent because they share a similar reflectivity to the surrounding surface, unlike the lunar maria which are less reflective and seem darker in appearance.

There is one other feature on Mercury which seems to be as common in the Solar System as craters – they have been seen on the Moon, Mars and Venus too. They generally appear as cliffs or some kind of fault but are now thought to be folds in the surface material and are known as 'rupes'. Their origin hails back to the evolution of the planet which, over many millions of years, has slowly cooled and contracted. With this contraction, the crust has had to fold and deform, leaving us with these rupes. However, until 2014 there was a problem: although the rupes suggested shrinkage there was no other direct evidence for Mercury actually having got smaller. Detailed observations by the Messenger space probe, which took high-resolution images of the entire surface, enabled accurate measurements to be made of the many surface features.

More importantly, the Sun was in a different position since the images were taken during the Mariner 10 mission; a different level of illumination meant that new features could be studied to build up a much more accurate profile of the surface. Messenger discovered new features that were the result of tectonic activity, like the scarps which are caused by faults in the crust that have pierced through the surface of the planet to altitudes of up to 3 kilometres, and the wonderfully named 'wrinkle ridges' that are smaller than the scarps but caused by the same tectonic activity. When compared to the rest of the Solar System planets, the rupes, scarps and ridges on Mercury seem to be particularly prominent, but when taken together the strong suggestion is that the planet has shrunk as it cooled in much the same way that the skin of an apple wrinkles as the fruit dries up. Studies indicate that the crust is on average 250 kilometres thick and that the planet has shrunk by around 14 kilometres since it formed over 4.5 billion years ago.

From the observations of Mariner 10 and Messenger, combined with Earth-based observation, it is possible to determine the density of Mercury by first working out its volume and its mass. Once we know its volume, which is easy enough to measure, we can determine its mass by studying its gravitational influence on objects in orbit around or flying close by it. This is relatively straightforward with planets like Jupiter, Saturn and even Earth which have moons in orbit around them, but Mercury has no such moon. Until the advent of space flight the only way we could estimate the mass of Mercury and indeed Venus, which also has no moon, was to measure the effect their gravity had on neighbouring planets. The changes in the orbits of other planets due to the gravity of Mercury are tiny so getting an accurate estimate of its mass was very difficult. But once space probes like Mariner 10 were able to pay a visit it became possible to measure how much the gravity of Mercury was tugging on the probe, thus revealing its mass. Dividing the mass by the volume tells us the density of the planet, and for Mercury this is 5.43 grams per cubic centimetre, making it the second densest planet in the Solar System, only slightly less dense than Earth. The high density of Earth can easily be attributed to the force of gravity compressing the core, but because Mercury is so much smaller, this cannot be the case. The only likely explanation is that Mercury must have a larger core that is rich in the heavier element iron.

Careful observation by spacecraft as they flew past has revealed clues about the planet's internal structure, and given such a high density it is believed that Mercury is broken up into the usual three zones. Below the crust is the mantle, estimated to be around 600 kilometres thick and made up of rocky silicates like the mantle of the Earth. The core, however, is much more interesting. Unlike Earth's, which is 17% of the volume of the planet, Mercury's core is a much more significant 42% of its body and is likely to be mostly molten.

There are two possible explanations why Mercury has such a high iron content in the core and the first looks back to the violent formation of the Solar System. Before planets existed, large chunks of rock known as planetesimals were commonplace and impacts took place between them, forming planets. It was one such impact on Earth that may have led to the formation of the Moon. The young Mercury is thought to have suffered a similar impact, and when it did, most of the crust and mantle is likely to have been ejected into space leaving behind the iron-rich core with substantially reduced outer layers. There are alternative theories, for example that the intense pressure from the Sun could have simply pushed away the lighter silicate rocks as Mercury was forming, or that the young

Sun was significantly hotter before it stabilized, causing the rocky surface to vaporize and get blown away by the onslaught of the solar wind.

The orbit of Mercury is highly elliptical, which means the planet is subjected to high tidal forces that keep the liquid core circulating. This movement is necessary to maintain the dynamo effect which is thought to drive the planet's global magnetic field. Studies by visiting spacecraft such as Messenger show that the magnetic field seems to be generally stable with the occasional vortex-like structure. These magnetic tornados are thought to form when the solar wind brings with it magnetic field lines from the Sun that connect up to Mercury's field and combine to produce a funnel. They are generally large in size, measuring upwards of 500 kilometres, and when they form, they expose the surface of the planet to the full force of the solar wind. The gaps in the magnetic field are not the only way the Sun can affect the surface of the planet but the magnetic field itself is strong enough that it can trap solar plasma, which can lead to weathering of surface features. It is also strong enough to have developed its own magnetosphere, which is a region of space surrounding Mercury where magnetic particles are controlled by the planet's magnetic field, as opposed to any other magnetic field – from the Sun for example – except where magnetic vortices form.

Swinging past Mercury on the way to Venus is a reminder of just how difficult it is to get a spacecraft to land on the planet, or enter into an orbit around it. The challenges stem from flying so close to the Sun, which exerts an immense pull of gravity. A good way of visualizing why this is a problem is to see the Sun as a large bowling ball sitting in the centre of a huge rubber sheet. It is easy to imagine that the presence of the ball will cause the rubber sheet to deform as it tries to support the weight. The dent it produces represents its gravitational pull, and as you 'roll' towards it, you fall deeper into it, making it harder to slow yourself down, just like running downhill. To get a spacecraft on to the surface of Mercury, or Venus for that matter, means flying into the 'gravitational well' of the Sun. That, without any effort on your part, makes you go faster, and the faster you go, the harder it is to slow down. In fact there are only a couple of ways to do it. One is to use aerobraking in the planet's atmosphere. This is fine for somewhere like Venus with its thick atmosphere, but Mercury is a different challenge. The only way to get a spacecraft in orbit around Mercury is to use rocket power to act against the acceleration from the force of gravity to slow the spacecraft and allow it to get captured by Mercury gently, rather than crash into it. Surprisingly perhaps, it actually takes more fuel to land on Mercury than it takes to get out of the Solar System.

Fortunately for us, we can use the RSU to drop down on to the surface and take a look around. With the Sun 'breathing' down our planetary neck it appears hugely dominating in the sky, almost three times larger than it appears from Earth and about seven times brighter – which, though it sounds a lot, isn't a particularly noticeable increase in brightness. The sky is still black, even at midday, because there is no atmosphere to scatter the light, and stars can be seen peppered against the blackness of space. In fact except for the fact that the Sun appears so much larger in the sky, you would be hard pushed to recognize any difference between standing on the Moon and standing on Mercury. You have landed in a region just to the north of the equator. As you survey the horizon you notice there is a lack of atmospheric extinction (in this sense referring to the extinction of light as it travels through the atmosphere) which, as on the Moon, makes it very difficult to determine distances to various features. On the northern horizon you can make out the shallow rim of Caloris Basin. As you scan around, you notice the craters are quite deep, and that is because of the lack of an atmosphere which on other terrestrial planets acts to slow down incoming lumps of rock, lessening the impact. Glancing down at your feet, you get your first proper look at the Mercurian regolith, which somewhat resembles what you saw on the Moon but with subtle differences. Its grain size seems to be finer than the regolith on the Moon, making it more powdery, so it sticks more readily to your boots. The reason for this is not fully understood but it is likely that the position of Mercury, further away from the asteroid belt and nearer to the Sun, could be the cause. Its distance from the asteroid belt suggests it may be subjected to more meteoroids of cometary origin which tend to have a higher velocity than those originating from the asteroid belt. Since they hold more energy, their impacts are more destructive, pounding the surface more on impact.

You cautiously make your way over to the rim of a nearby crater, moving in what can only be described as bounding leaps due to the lower gravity which makes this by far the most efficient way of moving around. Peering over the edge of the crater wall, you are surprised to see the floor buried deep in shadow. The crater is deep but the sides aren't steep, so after switching on the lamps attached to your helmet you rather cumbersomely make your way over the rim and down the slope. At the bottom of the crater the temperature is much colder than up on the surface, although of course your space suit is keeping your body at a comfortable temperature. Generally only in craters near the polar region does the temperature stay low enough for water ice to form; deep craters like this, closer to equatorial latitudes, tend to receive sunlight for a reasonable portion of their time which inhibits the formation of ice. Scanning around the crater floor, you follow the beam of your helmet light which happen upon the central peak of the crater, but it is very difficult to judge just how far away it is.

As you make your way back up the slope and out of the crater, the lower gravity and the finer regolith makes for slow progress, but eventually you are back up on the surface and before long back in the comfort and safety of the *Kaldi*, ready to head to your next planetary encounter.

At their closest, Mercury and Venus are just over 37 million kilometres apart, so the next leg of your journey will only take a few months and will be fairly uneventful.

On the approach to Venus, the second of the inner planets, the first thing you'll notice is its dazzling brightness. When viewed from Earth in the morning or evening twilight sky it appears as a bright star, outshining all others. But close observation, both from Earth and now as you make your approach, is rather uninspiring because hardly any features are visible. Like all objects in the Solar System except the Sun, Venus and the other planets are visible only because they reflect sunlight. Venus is a planet that is entirely covered in cloud from its north pole to its south pole and those clouds have a high albedo, which is to say they are highly reflective. This accounts for the planet's brightness – it's light bouncing off the cloud – and it also accounts for the fact that we can see nothing of the surface detail of the planet when we look at it.

Aside from the cloud, one of the less well understood atmospheric features of Venus is its ashen light. Because Venus is nearer to the Sun than the Earth it displays almost a full cycle of phases just like the Moon, sometimes appearing as a thin crescent and at other times almost full. The ashen light has been seen by only a few observers and appears as a subtle glow coming from the night-time hemisphere of the planet during some of the darker phases. Before the development of powerful telescopes there were some wonderful theories about its origin, one suggesting that Venusians were burning vegetation. A much more rational explanation is that it is similar to Earthshine seen on the night-time hemisphere of the Moon. Earthshine is the result of sunlight reflecting off the Earth and illuminating the darker part of the Moon; a similar process may be the cause of the ashen light. Another less popular idea says it might be the effect of several flashes of lightning happening in close succession, causing the atmosphere momentarily to glow, but a lack of radio emissions suggests this is not the case.

The thick, cloud-filled atmosphere of Venus makes visual observation of the planet's surface impossible. The only way to get a glimpse of what it is like is either to land or to use radar to pierce through the cloud. Radar is a technology that has many applications in space exploration but surface mapping is one of the most successful. The term radar is an acronym for 'radio detection and range' and it exploits the ability of solid objects to reflect radio waves, as demonstrated by Heinrich Hertz in 1886. Just eighteen years later, in 1904, Christian Hülsmeyer was granted a patent for the first radar device, the 'telemobiloscope', which bounced radio signals off distant unseen objects like ships in fog and revealed their direction and distance. It works because we know how fast radio waves travel (the same as all wavelengths of the electromagnetic spectrum, at 300,000 kilometres per second), and by timing how long it takes for the waves to be bounced back from an object we can determine distance.

This technology has been put to great use to pinpoint the

location of ships and aircraft but is equally important for mapping the surface of planets like Venus. Fortunately, radio waves will travel straight through the dense clouds and get reflected off the unseen surface, enabling us to build up a full topographical representation of it. Magellan used radar to map the surface of Venus and did so by sitting in a polar orbit, which means it orbited the planet from pole to pole. It was a fairly stable orbit so after each imaging run the planet would have rotated a little underneath, presenting a slightly different face. After completing thousands of orbits, each taking just over three hours, 95% of the surface had been mapped to a high resolution revealing many new features never seen before.

The Venusian atmosphere that Magellan peered through is composed almost entirely of carbon dioxide with small amounts of nitrogen. The high levels of carbon dioxide and the presence of thick clouds of sulphur dioxide are responsible for the intense temperatures felt on the surface, which were recorded by the Russian Venera probe: they reach 460 degrees – nearly five times hotter than boiling water. It is thought that at some point in the planet's history it was a much more hospitable place with an atmosphere more conducive to the existence of liquid water on the surface. However, over millions of years, evaporation of the water and volcanic activity have been responsible for the breakdown of the carbon cycle. On Earth, carbon is constantly being transferred between the atmosphere, animal and plant life, the oceans and the rocks. The process that allows carbon to get locked away in rocks is its interaction with water, either dissolving directly into the oceans at the surface and then being converted into organic carbon by living organisms – this ultimately finds its way to the floor of the ocean as creatures die and their carbon-rich skeletons become fossilized – or falling in rain, which causes weathering and allows some of the carbon to be absorbed by the rock. However it gets into the rock, it stays locked up for millions of years, or until one of the inhabitants of the planet decides to burn it as a fossil fuel. The lack of bodies of water on Venus means that the transfer of carbon from the atmosphere into the water and therefore rocks ceases, leaving it stuck in the atmosphere, and because carbon dioxide has the ability to absorb heat, it slowly warms up, leading to the greenhouse effect.

When the warmth from the Sun finally reaches the surface of Venus, Earth or any other planet with an atmosphere, it actually travels through the atmosphere without much interaction before arriving at the surface and slowly heating it. The surface then re-radiates energy at a slightly different wavelength which slowly warms the atmosphere. This is the driving force behind weather systems, where the warm air rises, creating areas of low pressure at the surface. The gentle warming of the atmosphere is checked on Earth because heat can slowly dissipate off into space, but this is not the case on Venus where the carbon-dioxide-rich atmosphere traps heat, preventing it from escaping, so the temperature rises unchecked. It is because of this runaway greenhouse effect that Venus is the hottest planet in the Solar System, hotter even than Mercury, which is many millions of kilometres closer to the Sun.

The dense atmosphere not only acts to cause an increase in temperature, but the low-level winds it generates tend to equalize night-time and daytime temperatures, so the surface is said to be isothermal. In other words, there is barely any diurnal variation. Even when temperatures at the poles are compared against temperatures around the equator there is little change. This is in stark contrast to Mercury where the daytime face roasts in the sunlight but plummets into frigidity by night. The only temperature variations experienced on Venus are seen with a change in altitude.

Temperature variations are the driving force behind wind, and with low temperatures at altitude and high temperatures at the surface there is a high temperature gradient between them. A high temperature gradient like this is partly to blame for the extremely high wind speeds seen at altitude, and on Venus they are getting stronger. When the Venus Express orbiter studied them it found that the average wind speeds at the top of the atmosphere, about 70 kilometres above the surface, were 290 kilometres per hour in 2006, but six years later they had increased to over 400 kilometres per hour. At this speed it takes a little under four days for the winds to blow around the planet, which is in stark contrast to the sedate rotation speed of the planet, which takes 243 Earth days to complete one revolution. A revolution in this case is retrograde or clockwise, just as it is on Uranus, whereas all other planets rotate anti-clockwise. The winds still confuse planetary meteorologists as they try to get to grips with Venus's extreme conditions.

The wind speeds can be measured by studying the movement of features in the clouds in the upper atmosphere. The majority of the carbon dioxide is concentrated in the lower levels of the atmosphere but above it are the thick clouds of sulphur dioxide that reveal the high wind speeds and reflect nearly 90% of the incoming sunlight. The combination of sunlight and sulphur dioxide is a deadly one for any living organisms that might find themselves on the planet's surface. Some of the incoming light gets absorbed by the carbon dioxide, sulphur dioxide and water vapour and triggers a chemical reaction. In particular, the ultraviolet portion of the sunlight causes carbon dioxide (one carbon atom with two oxygen atoms) to break down into carbon monoxide (one carbon atom with one oxygen atom) and one oxygen atom. The oxygen atoms that this process produces react with the sulphur dioxide to produce sulphur trioxide which, when combined with the trace levels of water vapour in the Venusian atmosphere, creates sulphuric acid.

The temperature in the upper levels of the atmosphere where this process takes place means that the sulphuric acid exists as a liquid and forms thick clouds along with the sulphur dioxide. The clouds, which tend to exist in a layer between 40 and 70 kilometres in altitude, produce sulphuric acid rain which then descends into the warmer lower layers. As the acid rain falls, it warms and releases the water vapour it is mixed with, making the acid more concentrated. By the time it reaches the surface, much of the sulphuric acid has dissociated into sulphur trioxide and water, both in their gaseous state. The sulphur trioxide dissociates further into sulphur dioxide and atomic oxygen which then combines with carbon monoxide to form carbon dioxide. The carbon dioxide tends to remain in the lower levels of the atmosphere but the sulphur dioxide and water rise through convection back to the upper layers of the atmosphere where the process begins again. The thick sulphur dioxide clouds are responsible for blocking most of the sunlight. Given that so much of the incoming light gets reflected back out into space, the surface of the planet is remarkably dark, even in daylight.

The first insight into the surface conditions on Venus came with the landing of the Venera probes in the 1970s. They returned the first images of the surface of this hostile world. On their descent, they measured that the sulphur dioxide clouds were about 40 kilometres thick but also detected hydrochloric acid and hydrofluoric acid, both of which are nasty chemicals. After finally touching down they revealed a world pretty much unfit for human exploration. Fortunately, of course, you can use the Reality Suspension Unit to explore the surface of Venus in complete safety.

Down on the surface it is dark, almost like constant twilight, and as a result of the thick clouds above you can't see any stars. Being able to stand and take in the scenery is itself a complete impossibility. The dense atmosphere above is exerting a pressure on you that is ninety times stronger than the pressure exerted by the Earth's atmosphere at the surface. That may not sound too much but it is the same as being a kilometre under water, and that immense pressure is enough to crush a human being. If that doesn't sound horrific already, then remember the surface temperature, which is around 460 degrees. So without the RSU you would be subjected not only to crushing pressures but also extreme heat. There is a bright side though, and that is that the extreme surface temperatures mean that the vicious sulphuric acid rain does not actually reach the surface, instead evaporating at an altitude of around 20 kilometres.

In addition to these harsh conditions, moving around on the surface is difficult, if not impossible and dangerous. The winds at the surface of Venus are moderately slow, rarely reaching anything quicker than a metre per second; compare that to the fastest wind speed on Earth of 113 metres per second, recorded during a tropical cyclone. Regardless of the sedate wind, the density of the atmosphere means it exerts quite a force on any object on the surface, which makes walking difficult. It also manages to lift and transport surface particles so it's like walking around in a sand storm, but instead of sand, stones and pebbles could be the projectiles.

Looking around at the surface reveals a world sculpted by volcanic activity, and it is apparent from the data gathered by Magellan that Venus has many more volcanoes than Earth. This is not because Venus is more volcanically active but its surface is much older than the surface of Earth. The crust on Venus is estimated to be about 600 million years old while the Earth's crust is only 100 million years old. The constant movement of the Earth's tectonic plates also means its surface is refreshed and renewed so any old volcanoes are likely to be slowly erased from view. Venus doesn't have plate tectonics so its surface is scarred by ancient volcanoes. Interestingly, volcanic activity on Venus seems to drive thunderstorm activity whereas on Earth we are used to storms being driven by rainfall instead. We have already seen that the only rainfall on Venus is of sulphuric acid but the extreme surface temperatures preclude the rain from reaching the lower levels of the atmosphere, so instead it is thought that Venusian storms are driven by ash particles that are being expelled during eruptions. The detection of lightning and thunderstorms on Venus suggests that some of the volcanoes are still active, and measurements of atmospheric sulphur dioxide also support the idea of recent eruptions because the quantities seem to have reduced between 1978 and 1986, which can be explained by a recent volcanic eruption sending plumes of the chemical high up into the atmosphere.

Like all the rocky objects in the Solar System, Venus displays thousands of craters, and the majority of them are still in excellent condition. This suggests that there has been minimal erosion of surface detail. More interestingly, it implies that the surface underwent some kind of global resurfacing event around 600 million years ago, but the lack of plate movement means the mantle is less able to redistribute and dissipate heat so instead it builds up until it reaches a critical level. In a global event that lasted perhaps up to 100 million years, the entire crust weakened and yielded to the mantle, in effect recycling itself.

We know that the surface is an inhospitable place because the various spacecraft that have landed there have demonstrated it. All of them were destroyed after just a few hours in the hostile environment. Spacecraft that orbited Venus were able to build up a gravitational map of the planet, allowing us to probe its inner structure. Common sense alone can help us to get a feel for the internal composition of the planet, which is similar in many ways to the Earth. Not only are they very similar in size, with Earth just 642 kilometres larger, but they have a similar density and mass. They also formed in roughly the same part of the Solar System, so it is a reasonable assumption that their internal structures resemble each other too.

The crust of Venus is similar to Earth's but with a uniform thickness. The crust of Earth varies from 10 kilometres thick for oceanic crust to around 50 kilometres for continental crust whereas the Venusian crust is believed to measure a fairly consistent 50 kilometres in depth, although the lack of moving plates suggests it may be even thicker than this. It also hints that the convection currents inside the mantle are less vigorous than those of Earth, whose plates are moved around by them. The mantle is thought to make up the majority of the bulk of Venus and mostly to consist of solid rock. Even though the mantle may be solid it acts like a viscous liquid – it can actually move, very slowly, over time. Convective currents within the planet are one of the primary ways in which heat is transferred from deep within to the surface. The heat is generated from the decay of radioactive elements in a process very similar to that on Earth. Even the temperature of the core is thought to be broadly the same, perhaps a little cooler, because a temperature that was much hotter would mean the mantle material would be less viscous, leading to more active convection and more geological evidence on the surface. Interestingly, the surface topography of Venus may be a direct result of the movement of the mantle. Where there are areas of higher than average surface levels there may be convective upwelling below, deforming the crust.

It is not only gravitational effects that allow us to probe inside a planet: we can tell a lot about the core of Venus by studying its magnetic field. The landers that visited the planet studied it extensively. One of the earliest, Venera 4, discovered it was much weaker than Earth's and that it was not generated by an internal dynamo like the Earth's. Instead it is the result of an interaction between the solar wind and the upper levels of the Venusian atmosphere, in particular the ionosphere. The different origin of the magnetic field tells us quite a bit about the core of the planet. For a dynamo to exist there needs to be three things: rotation at the core, a conducting liquid and convection. The Venusian core is likely to be made of the same chemicals as ours, which means it is composed of iron and nickel so it will also be conductive. If it is conductive and there is a very good chance it is rotating then the one ingredient missing is convection. For there to be an adequate level of convection there must be a liquid outer core like the Earth's, so no convection suggests that there cannot be a liquid core. The driving force behind such convection is usually that the bottom of the outer core is a lot hotter than the top and the high temperature difference drives the currents. The lack of a dynamo may alternatively mean that there is a liquid outer core but the lack of moving plates leaves no way for heat to escape. This may mean that the temperature at the top of the outer core is higher than it should be, thus lowering the temperature difference and reducing the convection currents. Until we can return to Venus with seismometers to probe the inner structure we will never really know the answer.

It's time to leave, because you have a seventeen-month leg ahead of you during which you will perform another orbit of the Sun, then slowly wind up speed by completing a gravitational fly-by of Venus again before taking in a final view of one of the most beautiful sights the Solar System has to offer, planet Earth. This will increase your velocity relative to the Sun by 21,000 kilometres per hour, and having attained that speed you can head out towards the outer planets, with Mars next on the itinerary.

FIVE

A Familiar World

YOU'VE LEFT BEHIND THE warmth of the inner Solar System and you're heading for an encounter with Mars, on average over 200 million kilometres away, making this the longest leg of the journey yet. And beyond Mars lies the asteroid belt, one of the most perilous parts of the trip. Is it possible to safely fly through it or is it like those scenes in science fiction movies where top pilots attempt to navigate through a veritable maelstrom of rocks? Time will tell. For now you must set out on the long haul away from the familiar surroundings of the inner Solar System.

During the months that will pass as you venture deeper into the Solar System there won't really be an awful lot to do – and those months will soon run into years of being cooped up inside your spacecraft. Things can get quite claustrophobic. On long trips like this crew selection is of paramount importance, because you want to be sure that if anything does go wrong, at least the crew will remain strong and working well together. It is for this reason that long-haul space flight planners favour couples because they already have a good bond and are used to living in close proximity to each other. Although of course sending couples into space means that before long, natural urges will take over and they will want to demonstrate their love for each other in a rather more physical way.

Until the twenty-first century the subject of sex in space was considered taboo among many space agencies, but with astronauts spending more time on board the International Space Station, researchers started to investigate the impact weightlessness would have on this fundamental human activity, not to mention its effect on procreation and pregnancy (there are social concerns too, as the development of intimate relationships within a small crew can lead to issues with performance and even the safety or ultimate success of a mission, but here we'll consider only the logistics of the physical act). For ships with artificial gravity it is not so much of an issue, but for those without it, sexual intercourse can be challenging. In the heat of passion, lovers will float around the room, often in opposite directions, leaving them floundering to get back to each other again.

In an effort to overcome the constant battle to stay together during moments of intimacy, the '2Suit' has been developed, with velcro and zips aplenty, its primary purpose to 'stabilize human proximity'. The suit allows its wearer either to attach himself to a work station to keep from floating away with minimal effort on his part or, and more appropriately, one 2Suit can be attached to another in such a way as to produce an almost cocoon-like garment, allowing its two occupants a chance for intimacy without unwanted interference from the laws of motion. And it's not just amorous couples who can benefit from such a garment: families with children would be able to sit close while watching a film, for example, not constantly having to exert themselves against drifting apart.

But for our species to become truly space-faring in the future we need to have a good understanding of the effect weightlessness has on the process of fertilization and pregnancy. Studies have already been undertaken using rats and mice to see how a foetus develops in these conditions. Fertilization of mouse embryos seems to have been successful in microgravity, although fertilization rates were lower than normal. Once the fertilized eggs were implanted into mice back on Earth, they seemed to develop into normal healthy mice. In different experiments, rat embryo development was studied. Everything seemed to progress normally until the rats, some of which were born in microgravity, were exposed to gravity back on Earth, at which point the microgravity rats lacked the ability to stand up. To date, no study has examined the full process of fertilization and foetal development from start to finish while in space, but it looks as though the main issues will be experienced once those born in weightless conditions return to a planet with normal gravity. More research is required. If we are ever to reach out among the stars and colonize distant planets then procreation will be an essential part of the process and will clearly be paramount to the survival of our species.

As this leg of the journey continues, the Sun slowly grows fainter behind you and the intensity of its light begins to drop. By the time you reach the vicinity of Mars it appears just over half the size you're used to seeing from Earth, and daylight on the surface of the red planet is similar to an overcast day back at home. The average distance from Mars to the Sun is about 228 million kilometres but its nearest approach takes it about 22 million kilometres closer. At that distance it takes just under 687 days to complete one full orbit of the Sun, which is about 1.88 times longer than an Earth orbit. Because of the orbital period of both planets, they experience a special alignment every two years and two months when the three bodies – the Sun, the Earth and Mars – line up, and this is known as 'opposition'. It gets its name because the Sun and Mars are opposite each other in the sky with the Earth in the middle. It's an alignment Earth enjoys with every one of the outer planets at some point. In the case of Mars, the distance between the two is usually about 90 million kilometres, and around the time of the next alignment there will be a flurry of spacecraft launched to head for the red planet.

An opposition of Mars means it is very well placed for observation from Earth, which is why it is without doubt one of the more popular planets to study among astronomers. This popularity goes back to the early days of telescopic activity around the 1600s. With basic and even crude telescopes, some very interesting observations of the red planet were recorded, and as the resolving power of telescopes improved some really quite remarkable detail could be seen. In September 1877, Giovanni Schiaparelli observed Mars with a 22-centimetre astronomical telescope during one of its particularly close oppositions and recorded a complex network of canals criss-crossing its surface. Schiaparelli actually gave the features the name *canali*, meaning 'grooves' in his native Italian, though the term was incorrectly translated at the time as 'canals'. These 'canals' were soon misinterpreted as a great global network of waterways transporting much-sought-after water from the polar regions to the drier, arid equatorial regions occupied by a race of evidently supreme intelligence. As telescopes continued to improve it transpired that the features Schiaparelli had seen were actually just an optical illusion. He was detecting surface features on the planet but the relatively low resolving power of his telescope meant the detail was not great, so his mind had tried to make sense of the markings by conjuring up lines joining them together.

The incorrect identification of canals on Mars may have fuelled the imagination of many who believed in a race of Martians inhabiting the red planet, but we now know that intelligent life does not exist on Mars. That is not to say, however, that some form of very primitive life could not exist. Already we have found evidence that at some point there was running water on the surface of Mars and that water molecules are locked up in the sub-surface layers of the planet and also in the north polar region. Back in 1984, a meteorite known as ALH84001 was discovered in Antarctica and it is thought to have come from Mars: scientists analysed the tiny pockets of gas locked up inside the rock and found them to be identical to the unique chemical composition of the Martian atmosphere, as analysed by the Viking missions that landed on the planet's surface in the late 1970s. Deep inside the meteorite are tiny little chains of chemical compounds known as polycyclic aromatic hydrocarbons (or PAHs for short) and these are thought to be the by-product of organic activity. So it seems that maybe, at least at some point in the past, there has been organic activity on the surface of Mars. But whether that activity ever led to the evolution of a more complex form of life like bacteria is still the subject of scientific debate.

Mars's red colour will no doubt be familiar to you, particularly if you have looked at it through a telescope from Earth. You should also be able to see the tiny specks of light from the two Martian moons, Phobos and Deimos, the discovery of which makes for quite an amusing story. They were first talked about by Jonathan

Swift in his famous book *Gulliver's Travels* just over 150 years before their actual discovery in 1877. The invention of the telescope in the early 1600s had revealed four moons in orbit around the mighty Jupiter so, knowing that the Earth had one Moon, Swift postulated that Mars would probably have two. Hardly scientific reasoning, but he did turn out to be correct.

Phobos is the larger of the two with a diameter of 22 kilometres compared to the 12.6-kilometre diameter of Deimos. It is thought that they may be captured asteroids as their composition is similar to bodies found in the asteroid belt. However, initially they would have entered into a highly elliptical orbit around Mars which contradicts the very circular orbits they both now have. It is likely that mechanisms such as aerobraking from the Martian atmosphere could have decreased their orbital speed which would have allowed their orbits to become more circular, but in the case of the smaller Deimos there simply has not been enough time for this to take place, so it is a mystery that remains unsolved.

On a journey like yours around the Solar System, if there is one place that really calls out for a visit, it's Mars. Walking on its surface and looking up at its two moons hanging in the sky is going to be an incredible experience. The planet itself is about half the size of Earth with a diameter of 6,779 kilometres but that does not stop people referring to it as Earth's twin. It is perhaps more appropriate to talk of Venus as Earth's twin, certainly based on its physical properties, but the conditions on the surface of Mars make it much more Earth-like.

As you approach, the white polar ice caps come into view. The atmosphere of Mars is mostly composed of carbon dioxide – similar to Venus, although it is much less dense so the surface temperature is nowhere near as high. Compared to Earth, the average surface pressure on Mars is about 0.6% so it would not sustain life and you'll need to explore with the protection of a space suit. The suit will maintain an appropriate atmospheric pressure not only to aid breathing but also to ensure your body does not balloon up and make movement difficult. It achieves this by carefully pumping pure oxygen into its gas-tight confines.

From the surface, the moons are actually a little bit of a disappointment. Deimos is no more impressive than Venus when viewed from the Earth, but Phobos is a little more interesting. It appears about a third the size of the full Moon in the sky above the Earth and because its orbit around Mars is aligned with the Martian equator, it appears quite low in the sky from your landing spot just north of Hellas Basin in the southern hemisphere. If you had landed around the north or south polar regions, then Phobos would have been completely out of view. From here, though, both moons can be seen.

Observation of them over just a few hours reveals a startling difference between their orbital characteristics. Deimos is the outermost of the two moons and takes about thirty hours to complete one orbit, but when the rotation of Mars is taken into account, it takes about 2.7 days to slowly track across the sky from east to west. Phobos, on the other hand, is a little more nippy and orbits Mars so fast that it appears to rise in the west and set in the east with just seven hours between successive rises.

Both Phobos and Deimos present phases much like Earth's Moon, although it requires a telescope to see the phases of Deimos because of its changing brightness. Phobos is much easier to see and, like the Moon, is tidally locked, meaning that it constantly presents the same face to Mars. Due to the proximity of Phobos and its fast orbital speed, tidal forces between it and Mars are actually beginning slowly to decelerate it in its orbit, which will eventually make it fall towards the surface. Before it crashes into the planet, though, it will reach the Roche limit where the tidal forces become so strong that they will rip the moon apart. There are a number of chain craters on Mars which are thought to have been caused by other young small moons that suffered such a fate. Deimos, on the other hand, is a little too far away and instead is being accelerated away from Mars.

Surveying the scene on the surface of Mars will no doubt remind you of the images sent back by the Viking landers. It feels like you're in the middle of a vast red desert. Standing still is OK but moving around is challenging because of the lower gravity. It's easy to lose your footing. The planet is about a tenth of the mass of Earth which means the force of gravity is just under 40% of what it is on Earth, so if you stepped on a set of scales on Mars you would weigh under half your normal body weight. The surface is not too dissimilar to the Moon's with a powdery coating almost the consistency of talcum powder sitting on a basaltic rock upper crust, but there is one big difference: everything looks a salmony-red colour. The fine powdery coating is iron oxide, more commonly known as rust, exactly what is found on Earth on iron objects that have been left outside in the rain. On Mars the iron oxide formed billions of years ago when iron reacted with the then more plentiful supply of liquid water.

The water that existed on Mars billions of years ago also reacted with the carbon dioxide in the atmosphere and produced carbonate rocks, a process that extracted carbon dioxide from the atmosphere and then locked it away, slowly thinning it. The lack of tectonic activity on the planet means the carbon dioxide has remained locked away inside the rocks, leaving the Martian climate very different to how it was billions of years ago. The thin atmosphere and highly elliptical orbit of the planet are responsible for the pretty high temperature differences across Mars. Here on the rim of Hellas Basin at about 30 degrees south of the equator the midday temperature hits highs of about 10 degrees, but at night it plummets to around minus 60. The temperatures across the rest of the globe also vary widely, with highs of around 20 degrees in the northern hemisphere during the summer down to minus 150 degrees in the polar winters. The substantial temperature differences between the two hemispheres drive some quite powerful winds. The Viking landers recorded speeds of 30 metres per second (about 108 kilometres per hour) but because of the low atmospheric pressure the winds would not feel as strong as the same wind speed on Earth.

These winds were responsible for distributing the iron oxide around the planet and are still responsible for the global dust storms that engulf it, such as the event in 2001 which was monitored by the Mars Global Surveyor spacecraft, which was in orbit at the time. The fine nature of the dust that covers the surface is responsible for the rather strange and eerie pink sky too, because the individual particles are small enough for many to get left suspended in the thin Martian atmosphere. When the first pictures were received from the Viking probes they were corrected by the people who received them to show a blue sky. It was soon realized that the colour cards fixed to the side of the craft that allowed for a correct colour balance to be applied were now showing the wrong shades. Once they were corrected, a true colour image was seen and for the first time the alien pink sky of Mars was revealed.

The dust storms that often plague the planet are a real challenge for surface rovers and landers because they can easily deposit a fine layer on solar panels and optical instruments rendering them inoperative. As a direct result of the surface conditions, Mars landers have batteries that charge from solar panels to keep the power topped up during periods of reduced incident lighting when dust storms strike. Fortunately for you there is no dust storm under way today but the fine powdery nature of the dust looks like it could cause problems for ventilation systems. Certainly trying to survive the Martian surface without a space suit would more than likely be suicidal, not just because of the lack of air pressure, oxygen and extreme temperatures, but because the dust would, if unfiltered, lead to suffocation.

Dust storms seem to originate from certain areas of the planet and one such location is the Hellas Basin, which is the chief reason why this spot was chosen for your flight plan. This large impact crater measures a staggering 2,300 kilometres from side to side and is 7 kilometres deep. It's the largest visible crater in the entire Solar System. At this depth, the atmospheric pressure rises to about 0.01 atm (Earth atmospheres), which is low but still higher than the average surface pressure. If the temperature could rise above about zero degrees then it is just possible that liquid water could exist on the surface. It may be responsible for some of the erosion features that have been seen, such as the many gullies that run to the north-east.

The similarities between Earth and Mars and its relative proximity to us make it the prime target for possible human colonization. In its current state, Mars is a great place for a short excursion to the surface but a bit more planning and preparation would be required before a human base could be properly established. Grand plans of terraforming Mars to make its surface capable of supporting life remain a very long way off, not to mention morally questionable: our exploration of the planet has so far shown up no conclusive signs of life, but until we can categorically state that Mars is completely devoid of any life, however primitive, then it would be wrong potentially to put any life forms at risk. But certainly setting up a human base in specially designed habitats to support life is within our grasp.

To make that vision a reality it would be prudent first to send some robotic missions to the planet to start transferring equipment and materials. The first few shipments could include inflatable habitation units and electricity production technology such as solar- or wind-driven units, along with the ability to store the energy created. Machinery and tools would also be needed to extract materials and minerals from the ground for production of food and drink, and later for building materials. There would also need to be equipment that could extract chemicals from the atmosphere to produce breathable air and rocket fuel. This is just the tip of the iceberg though.

Once enough infrastructure had been sent over and established as much as possible, the first human settlers would arrive. They would have to be people with a proven record of being able to live together, so again, it's likely that couples will be the first to be sent. They would also need to be entirely self-sufficient so they would have to have a wide skill set, not only from a construction, engineering, medical and scientific perspective, but as the colony grows there will be a need for some form of governing process too, to ensure law and order. In essence, a new society would have to be created, one whose participants will probably have to be hand-picked initially in order to give it the best chance of success. Longer term, such a colony would need an injection of people with more diverse skills to help the new society to flourish.

For now, though, your time on the planet has come to an end, and you are leaving behind what remains a desolate, barren world.

From the comfortable surroundings of the *Kaldi* you can once again look down on Mars which, like Earth and many other planets, has an axis of rotation which is tilted with respect to the plane of its orbit around the Sun. The axial tilt of Mars is 25 degrees which makes it very nearly the same as the Earth's, which at 23.5 degrees is a little less. More crucially, and as a direct result of this tilt, Mars experiences seasons just like Earth, although there are some fairly significant differences. The orbit of Mars is just under two Earth years long, so the seasons are generally about twice as long. Mars also has a very eccentric orbit which means that its distance from the Sun varies by as much as 19%, so the seasons are not the same between the two hemispheres, unlike on Earth whose seasons follow the same but juxtaposed pattern in the northern and southern hemispheres. The seasons in the northern hemisphere of Mars bring temperatures that are on average about 30 degrees cooler than the same season in the southern hemisphere, but this has not always been the case. There is strong evidence that the tilt of Mars has changed over its history, having been much more extreme in the past. Deep under the surface there are vast reservoirs of water locked up in ice which are believed to be the result of a much larger polar ice cap that in previous millennia extended to more temperate latitudes on the planet. As the tilt reduced to the present value, the polar caps receded, leaving any evidence of their existence locked away underground.

An exciting discovery was made back in 2013 by the Opportunity rover which found evidence of neutral water on the planet's surface. Locked up inside some Martian rocks the Opportunity found clay minerals that could not have formed from Mars's rather more acidic version of water. Neutral water is close to what we would consider to be drinking water, so the evidence tantalizingly suggests that conditions on Mars a few billion years ago were conducive to the existence of life.

The polar ice caps of Mars are once again clearly visible from your vantage point, the southern cap looking smaller than the northern, so it must be summer in that part of the planet. The northern hemisphere is pointing away from the Sun causing it to experience winter, and as a result that polar region is plunged into permanent darkness. The drop in temperature causes carbon dioxide in the atmosphere to desublimate (a process where gas turns directly into a solid without going through its liquid phase first) into solid chunks of carbon dioxide ice that cap the polar regions. The ice caps have significant amounts of water ice too but these are generally covered by a layer of carbon dioxide ice, and together they form a polar cap that is between 2 and 3 kilometres thick. If all the water ice on Mars thawed, including that at the poles, there would be enough to cover the planet in a sea nearly 10 metres deep. Detailed images of the polar regions reveal strange almost spiral-shaped troughs in the ice which are sculpted by polar winds. This is a wonderful example of the Coriolis effect, which causes air to be deflected due to the rotation of the planet and is the same reason why the air circulates around regions of high and low pressure on Earth. With the change in the seasons and the onset of spring and summer, the poles warm, allowing the ice to sublimate back into the atmosphere, generating winds that blow from the poles.

There are many other large features that are prominent on the surface of Mars. The most stunning is Valles Marineris ('Mariner's Valley'), named after the Mariner 9 spacecraft that discovered it. It is a vast canyon system that runs for 4,000 kilometres across the Martian surface, at its widest part spanning 200 kilometres and descending 7 kilometres down into the Martian crust. Comparing those statistics to the Grand Canyon in Arizona, which is 446 kilometres long, 29 kilometres wide and 2 kilometres deep, makes you realize just how large it is. There are many cracks and fissures that run off the main valley, which is classed as a typical (albeit very big) rift valley. This type of feature is common in the Solar System. There are similar examples on Earth and on Venus, where a linear-shaped feature runs between highlands or mountain ranges and is the result of some kind of geological fault. The valley is thought to be the result of tectonic activity that appeared as Mars cooled early in its history. The crack then widened as the crust in the Tharsis region to the west started to rise and erosional forces began their assault. Closer inspection shows that some of the channels on the eastern side appear to have been carved by the erosive forces of running water.

The Tharsis region itself is a huge volcanic plateau that is home to three enormous shield volcanoes (domed-shaped volcanoes with gently sloping sides): Arsia Mons, Pavonis Mons and Ascraeus Mons. Just on the edge of the plateau is the tallest volcano in the Solar System, Olympus Mons, which towers 22 kilometres above the surface and dwarfs Earth's largest volcano, Mauna Loa on the island of Hawaii, by 16 kilometres. The plateau has developed because it is over a hotspot in the underlying mantle known as a superplume, where vast quantities of hot, dense magma well up, forcing the crust to rise. The lack of individual plates on Mars means that this magma has built up over billions of years, unable to go anywhere and producing huge volcano complexes. Unfortunately there is unlikely to be a major eruption as a considerable quantity of the magma will slowly cool and solidify. It was once thought that the three volcanoes present on the Tharsis Plateau were individual structures but it is now believed that they represent one complex system – a theory which is supported by the many geological features that surround the region.

Olympus Mons is very much a separate system, but despite its large size it is actually pretty unimpressive from the surface. Being a shield volcano, the slope of its flanks is no more than about 5 degrees, so had you stood at the foot of this impressive feature you would actually have been unaware of it. The size of Mars too would have precluded any great views of Olympus Mons as the horizon is only 3 kilometres away and the extent of the volcano takes it beyond the visible horizon. Even standing on top of it you would still have been unaware of the nature of the feature below you as the ground would have just gently sloped down away from you.

What you would have experienced at the summit, though, is clouds. The atmosphere at the altitude of the peak is only 10% of the surface pressure but even so, the flow of air across the peak is sufficient to cause orographic uplift. The process by which clouds usually form begins with a parcel of air which sits on the surface and contains an amount of water in its gaseous form. Heating from the surface causes the cloud to rise and cool until it reaches its dew point when it can no longer hold the water gas. At this point, the air is saturated and water condenses out to visible droplets that we see as cloud. Orographic uplift is slightly different as it is the profile of the terrain that forces the air to rise – for example, air flowing across Olympus Mons gets forced up and cools, allowing cloud to form at the top. This process is often seen over mountains on Earth too and can be a real danger for climbers who suddenly find themselves engulfed in thick cloud.

The incredible size of Olympus Mons is largely due to the fact that there is no tectonic activity on Mars. A volcanic vent that sits over a hotspot will stay there, allowing molten lava to flow out, and it is very likely that the volcano's sheer size is the cumulative effect of countless lava flows over billions of years that have simply built up. Unfortunately it is very unlikely that a lander will be sent to Olympus Mons because the atmospheric density at the summit is far too low to allow for parachutes to slow a lander's descent. That said, a more complex landing technique could be employed like that demonstrated by the Mars Science Laboratory mission in 2012, where the lander was lowered on a winch from a rocket-powered hovering platform. For now we will have to be content with images and data gained from Martian orbit, and fortunately we can still learn a lot by using these gravitational mapping techniques.

Images returned from the MSL show the flanks of the volcano to an incredible level of detail; intricate grooves, channels and ridges that have been forged by the lava flows can be seen. By studying the distribution of craters it is possible to age the surface to anything between 2 and 100 million years, which in geological terms is very recent, suggesting that Olympus Mons is still active but only in a very sedate fashion. The caldera of the volcano is made up of six overlapping craters, but not craters formed by impact: instead they are the result of surface collapse. As the molten magma seeps out of the volcano through its many vents, the surface becomes unsupported and collapses at the top. Therefore each caldera at the top of Olympus Mons represents one of the major expulsion events. In much the same way that the study of craters on the Moon can help us to determine roughly the age of the lunar surface, by studying these caldera it is possible to determine that the largest magma chamber lies about 32 kilometres under the surface. Each of the six caldera are thought to have formed within a million years of each other and are probably between 150 and 350 million years old.

All over the planet, from the Tharsis Plateau to the Valles

Marineris, there is evidence of geological activity. The surface is likely to be one large plate, which accounts for the many surface features shown, although there is an alternative theory suggesting that Valles Marineris may be the result of two plates slowly moving apart. The crust itself is about 50 kilometres thick and composed mostly of silicon and oxygen which are locked up in silicate rock, but there are quantities of iron, magnesium, calcium and potassium. Deep underneath the crust are two concentric zones that represent the mantle and core of the planet. Differentiation is the process that drives the production of the very distinct zones on Mars and other rocky planets, and it's a process that occurs because of the different behaviour and properties of the material. The mantle sits below the crust and is the region responsible for the production of many of the Martian geological features we can see today. Below that is the partially molten core which is thought to be 3,500 kilometres in diameter and made of iron and nickel with small amounts of sulphur.

But the most tantalizing aspect of Mars remains its potential suitability for long-term human habitation. Clearly at the moment it cannot support human life, chiefly because of the low atmospheric pressure, but extensive research into the possibility of humans inhabiting foreign worlds using artificial environments may just make it achievable. Experimentation into transforming the atmosphere of Mars has received serious consideration, but while this process is technically feasible it will take thousands if not millions of years to complete.

Should we ever manage to establish a human colony on Mars then something as seemingly simple as communication with Earth will be a little more problematic than you might think owing to the immense distances between the two planets. Any communications will be sent via radio waves and would therefore travel at the speed of light, which is 300,000 kilometres per second. At that speed it would take just three minutes for a message to traverse the space between the two planets when at their closest, or twenty-two minutes when they are furthest apart. When the two are on opposite sides of the Sun, communication will not be possible unless a satellite is placed some distance from Earth to relay the signal.

But still, there is a lot to be said for choosing Mars as a possible human outpost: the day is about the same length as an Earth day, a year is only about 320 days longer than Earth's, and the seasons are broadly the same even though they are longer-lasting. A number of challenges will need to be met, but these could be overcome through the use of an artificial and self-contained environment on the surface of Mars inside which humans could live. Such environments have been tested on Earth already, like the Mars 500 mission where volunteers were locked up in a mini self-contained ecosystem to see what physical and emotional effects it had on them. It is quite possible to construct a habitable living environment on Mars just like this, and from there the first inhabitants would be able to explore and could slowly build a new human society.

An outpost on Mars would provide a great base from which missions could explore the outer Solar System. From there, not only are the large gas giants within reach but also the asteroid belt, which is your next destination. Since flying past the Earth on the way to the outer reaches of the Solar System we have travelled 100 million kilometres to get to Mars; to get to the inner boundary of the asteroid belt will take a further 100 million kilometres.

Before the discovery of the first asteroid, in 1766 the German astronomer Johann Titius scribbled a note suggesting that there was a numerical pattern in the distances between the planets. He realized that if you start off with the sequence of numbers 0, 3, 6, 12, 24, 48, 96 then add four to each and divide by ten, you end up with the following series: 0.4, 0.7, 1, 1.6, 2.8, 5.2 and 10. At first glance those numbers do not look too significant, but compare them to the average distance between the planets and the Sun in astronomical units (where 1AU is the average distance between the Earth and Sun) and you will find that Mercury is at 0.38AU, Venus at 0.72AU, Earth at 1AU, Mars at 1.5AU, Jupiter at 5.2AU, and finally Saturn is at 9.5AU. In 1768, another German astronomer, Johann Bode, referenced this number sequence in one of his publications but did not give credit to Titius so it became known as Bode's Law. What made this series even more startling was that the discovery of Uranus in 1781 showed its average distance from the Sun to be 19.2AU, which almost matches the next number in the sequence, 19.6.

Look back at the numbers, though, and you'll notice that there appears to be a gap. Mars has an average distance from the Sun of 1.5AU and Jupiter is at 5.2AU, but in the Titius–Bode sequence there is a number between them, 2.8. Titius even raised the question about another as yet undiscovered planet in orbit between Mars and Jupiter, and following the discovery of Uranus, in 1800 a team of astronomers led by Franz Xaver von Zach started to scour the sky in search of the missing planet. They were each allocated a portion of sky that spanned 15 degrees along the path that all the planets seemed to follow. In January 1801 Giuseppe Piazzi announced the discovery of a tiny moving object in the orbit predicted by Bode's Law. Telescopic observation showed that it was a fast-moving object with a motion similar to that of the comets, although a lack of the fuzzy coma that surrounds comets led them to believe it was indeed a planet. Even under high magnification it was impossible to resolve the tiny object into a disc so it was only its motion that seemed to suggest it was any different from the surrounding stars. It was given the name Ceres after the Roman goddess of the harvest, and just over a year later, in March 1802, the discovery of a second object was announced and called Pallas. They became known as asteroids, from the Greek *asteroeidas*, which translates as 'star-like'. Both share a very similar average distance from the Sun of 414 million kilometres, or 2.7 astronomical units.

The new asteroids seemed to be orbiting the Sun at the right distance to fill in the gap in Bode's Law, but the law has received much scrutiny over the years and it has been concluded that the sequence is actually nothing more than a mathematical coincidence. The discovery of Neptune in 1846 at a distance of 30.1 astronomical units compared to the Titius–Bode number of 38.8 further discredited the concept, suggesting coincidence rather than a real phenomenon.

Furthermore, Ceres and Pallas turned out to be just two of millions of asteroids that orbit the Sun between Mars and Jupiter. It is estimated that there are around 200 larger than 100 kilometres, nearly 750,000 that are larger than 1 kilometre, and perhaps millions that are smaller fragments or pieces of dust. Ceres is the largest of them all with a diameter of 950 kilometres, and Pallas the second largest at around 550 kilometres in diameter. When added to the mass of the third and fourth biggest, Vesta and Hygiea, the four of them account for half of the mass of the entire asteroid belt.

How this asteroid belt formed has been the cause of much controversy over the years, with theories ranging from exploding planets to planets that have been destroyed by cometary collisions, but the modern accepted theory is a little less dramatic. We have already seen how the Solar System formed when a large cloud of gas and dust slowly collapsed under the force of gravity, within which the temperatures and pressures became so extreme that nuclear fusion started to occur, giving birth to the Sun. Around the hot young Sun was an accretion disc of dust within which a number of random collisions took place, some of which caused the chunks to stick together. As the collisions continued and the chunks grew in both size and mass they started to become gravitationally dominant, pulling other chunks of rock towards them. Over time, the rocky inner planets and the gaseous outer planets formed, but between them another planet was struggling to form. The presence of the large and very dominant planet Jupiter prevented the further growth of any planetesimals in the belt and instead the remaining group of rocks continued to orbit the Sun together.

As the Solar System evolved and the majority of the planets slowly migrated inward from their initial positions, the increasing strength of Jupiter's gravity on the asteroid belt caused many of them to accelerate as it adjusted their orbits. In most cases, this was the result of asteroids experiencing orbital resonances with Jupiter. Orbital resonances are not an unusual phenomenon in the Solar System and they exist where two or more bodies exert a regular gravitational interaction which over time has a cumulative effect on both of them. A good analogy to this is a parent pushing a child on a swing: regular pushing with the same force at a constant frequency will lead to the swing going higher and higher. It's the same with objects in the asteroid belt and Jupiter: the regular tug experienced by the asteroid as it passes by Jupiter on each orbit will act to adjust the speed and therefore the orbit of the asteroid until it is in resonance with the orbit of Jupiter – for example an asteroid might complete two orbits for every one of Jupiter's. In the majority of cases this is an unstable situation, and the continued gravitational interaction will eventually lead to the orbital resonance breaking down again.

This influence from Jupiter is also responsible for reducing the overall mass of the asteroid belt early in its history by a factor of a thousand. At the current observed inner boundary of the belt, just over 2AU from the Sun, an object in orbit will soon fall into a 4:1 resonance with Jupiter, completing four orbits for every one of Jupiter. With the cumulative effect of the regular tugging on the asteroid it will eventually become ejected from the belt into a new orbit outside the belt. Any asteroids that wander too close to the Sun will get swept up by the gravitational pull of Mars, whose furthest point in its orbit takes it to around 1.67AU. While Jupiter is responsible for inhibiting the growth of a planet within the belt, Mars and Jupiter together keep the components of the belt in check. If they try to wander too close or too far, they will either get ejected or put back in their place. A similar process takes place around Saturn where tiny moons in orbit around the planet keep particles of ice and dust in place, creating the majestic ring system.

The Sun has also influenced the development of the asteroid belt and the asteroids within it. During its early more formative years the temperature in the young Solar System would have been higher, and some of the larger asteroids would have partially melted. As they became molten, the heavier elements would have sunk and the lighter elements risen, leading to a body that is differentiated. It is quite likely that some would have experienced volcanic activity, with huge lakes of molten lava forming and then solidifying over millions of years. At the other end of the spectrum, any asteroids that formed at a distance greater than 2.7AU would have accumulated ices because at that distance from the Sun the temperature dropped sufficiently for water to freeze. In 2006, the discovery of a number of cometary nuclei in the outer reaches of the asteroid belt was announced, and it is thought that some of these may well have brought water molecules to the young Earth as it was forming, and that ultimately led to the existence of our oceans.

The asteroids within the belt broadly fall into three categories based on their composition. Around the inner region of the belt they tend to be rich in silicates, which is a particular type of rock composed of varying quantities of silicon and oxygen. Silicates are common in the inner Solar System and make up about 70% of the crust of the Earth. Silicate rocks are generally the result of geological processes such as partial melting or crystallization that have modified the original protoplanetary disc material out of which the asteroids formed. These types of asteroid are known as S-type asteroids, and many of them are also found to have traces of metals and carbon-based compounds. This is quite a contrast to the carbonaceous asteroids which are found around the outer rim of the belt, which as their name suggests are rich in carbon-based compounds. Unlike the S-type asteroids their composition is thought to be much more representative of the composition of the early Solar System. They make up the vast proportion of asteroids within the belt, numbering well over 70%, but though they are plentiful they are the hardest to see because of their low reflectivity.

The final common type of asteroid found within the belt is the metallic or M-type; as the name suggests, they are rich in metals like iron or nickel. They are fewer in number when compared to the other two types, making up only 10% of the overall population of asteroids in the belt, and their composition suggests that they originated from the core of differentiated asteroids, where the heavy metallic elements had settled to the core and been subsequently liberated by some form of collision. The theory of differentiation among the larger asteroids also predicts that they should form crusts composed mostly of basaltic rock as the lava from the partially molten asteroid rapidly cools. This in turn suggests that there should be much higher quantities of basaltic rock and basaltic asteroids in the belt than have been observed to date.

With the composition of the asteroids so well understood, an exciting possibility exists to mine them for minerals, either to bring back to Earth or to use for the construction and running of space exploration and colonization activities. Theoretically, at least, iron, nickel, titanium, oxygen and hydrogen could all be mined from asteroids within the main belt, or indeed from any asteroids wandering around the Solar System. Whether they should be used to replenish the dwindling and finite reserves on Earth, which are forecast to run out by 2100, is an ethical dilemma, and an operation that would come at an enormous financial cost, but the idea of using them as supplies for space exploration seems to have received much more support. It is not just the asteroids, though: there are thought to be many exhausted, almost extinct cometary nuclei which could be mined for oxygen by passing spacecraft and which could be used to produce rocket fuel or air to breathe, while the heavier metals from the asteroids could be used to build spacecraft in space, or repair damaged craft. The possibilities are endless.

Choosing which asteroids to mine would be the first challenge to overcome, regardless of the intended use of the mined material. One of the first considerations is the location and orbital parameters of the asteroid, because put simply, some are easier to get to than others. The considerations will be different depending on the point of destination. For example, when travelling through space on a voyage such as yours the velocity of the spacecraft is likely to be high so it may take a considerable amount of fuel to be able to adjust the velocity sufficiently to match that of the asteroid. Expeditions from Earth, however, need to make sure that travel to the asteroid can be achieved using the Hohmann Transfer orbit, which is among the most fuel-efficient ways of travelling to a destination. During this type of journey there are two bursts from the rocket engine, one to increase the velocity of the spacecraft and send it into a highly elliptical orbit to intercept the destination, and another to move it off the transfer orbit and on to an orbit around the Sun that matches the orbit of the destination. The process can be reversed to get back to Earth but this time by firing the rocket engines in the opposite direction to slow the spacecraft, first to return it on to the Hohmann Transfer orbit and again on arrival at Earth to drop into Earth orbit. The concept of the Hohmann Transfer orbit is pretty straightforward and was first discussed in 1925 by the German scientist Walter Hohmann, after whom it is named.

How essential it is to plan the spacecraft's trajectory was beautifully demonstrated with the Rosetta mission that arrived at Comet 67P/Churyumov/Gerasimenko in 2014: ten years were spent carefully adjusting its trajectory in order to match its orbit around the Sun with that of the comet. The huge effort was worthwhile, though, as it meant the probe was travelling around the Sun at the same speed as the comet so that the relative speed between the two was as low as it could be, which made the landing so much easier.

So we know we can do it, and all the types of asteroid we have looked at offer something in terms of mining activities. The C-type carbonaceous asteroids have high abundances of water which could be used for life support; the oxygen and hydrogen molecules could also be dissociated to provide rocket fuel. By their very nature they are high in organic compounds too which can be used for production of fertilizers to help with food production. The S-type asteroids contain very little water or organic compounds but are rich in metals including iron, nickel and even gold and platinum. The M-type asteroids are few and far between but are worth hunting down as they contain up to twenty times more metal than the S-types, making them a great source of material for repairing equipment or even building new spacecraft.

Deciding which asteroids to mine and getting to them is one thing; actually doing the mining is a whole different challenge. A number of different approaches have been described in various works of science fiction. In the film *Alien*, the crew of the mining spaceship *Nostromo* returned to Earth after a specific mission with almost 20 million tonnes of material. In reality, acquiring such a haul from our own asteroid belt would be a tall order. Even if you were to capture the largest of the asteroids and drag it all the way back to Earth you would only gain about 1,000 tonnes of material, and that would still need to be refined. The practicalities of mining asteroids are a little different from what we see in the movies.

Perhaps the biggest challenge is the lack of a significant gravitational field to hold spacecraft, equipment and even astronauts on the surface of asteroids. Unlike the Moon, where astronauts were able to walk or hop around, if you tried that on any of the asteroids you would more than likely just float off into space. It would be necessary to use hooks and tethers to attach anything that needs to stay where it is in order to stop it floating off. As a prelude to such an exercise, NASA have been training astronauts under the surface of the Atlantic Ocean, not only to simulate the experience but also to test techniques and equipment they might use.

This may sound pretty straightforward, but as the European Space Agency found out when they tried to drop the Philae lander from the Rosetta probe, the process is fraught with difficulties. After touchdown on the comet nucleus, a jet was supposed to fire to push the lander against the surface while some screws worked their way in to attach the craft; at the same time, harpoons should have fired to secure it firmly to the surface. Both the jet and the harpoons failed, leaving Philae only tenuously attached to the surface, just one false move needed to send the tiny little craft careering off into space again.

Assuming it is possible to attach the members of a mission to the surface, several different mining approaches could be employed. Many of the asteroids are thought to have rubble-like surfaces so a simple scoop or claw might be used to scrape off the surface material. Those with a high metal content like the M-type asteroids may be covered in metallic grains or granules which could be harvested with a magnet that sweeps over the surface. Perhaps the most adventurous and challenging approach, though, is to dig a mine shaft and extract minerals from deep within the asteroid. This is by far the most technically complex operation, not just from an engineering and logistics point of view but also in terms of working out exactly where to drill in order to get to the minerals.

With millions of asteroids to choose from you would think that finding a suitable one would be an easy task, but remember, the four largest asteroids account for nearly half of the mass of the entire belt. Identifying the exact mass of an asteroid is done in much the same way as for other objects in the Solar System, through the examination of their gravitational interaction. In the case of Ceres, the largest of the asteroids, it weighs in at 940 billion billion kilograms. That does sound quite a lot, but when you think that it would take twenty-five of them to equal the mass of the Moon, which is itself quite small, that puts the figure into perspective.

It measures just under 1,000 kilometres in diameter and because of its large mass it has achieved hydrostatic equilibrium, which you will recall means it is roughly spherical in shape. (Even though this is one of the criteria for the definition of a planet as laid down by the International Astronomical Union, Ceres has not become gravitationally dominant in its orbit so it is officially classed as a dwarf planet with all other objects in the belt classified as asteroids.)

Unlike the planets we have looked at so far, Ceres is unusual because its body is not differentiated, which means the metals have not had a chance to separate from the rocks. The core is believed to be a rocky composition and observations suggest this may be surrounded by an icy mantle. Spectroscopic studies have identified nearly 400 million trillion gallons of water ice in the mantle of Ceres compared to an estimated 326 million trillion gallons on Earth. There is a problem with this theory, however, because Ceres presents a rocky surface, and a rocky surface overlaying an icy mantle is an unstable situation as the force of gravity would attempt to drag the rock down through the ice. This would leave significant salt deposits on the surface which have so far remained elusive to spectroscopic studies. There does seem to be a very tenuous atmosphere on Ceres, and although there is some evidence of water ice on the rocky surface, the low atmospheric pressure means that the water soon turns into a gas through sublimation and escapes into space.

Pallas was the second asteroid to be discovered, and with an average diameter of 544 kilometres is generally referred to as the second largest of the known asteroids. Vesta is a very similar size with an average diameter of 525 kilometres, and their irregular shapes mean there is often contention over which is larger. Pallas just about beats Vesta on size, but when we compare the mass of the two, Pallas is about 30% less massive. Its composition somewhat resembles some of the meteorites that have been found on the surface of the Earth which are high in carbon compounds, and spectroscopic studies suggest a surface with high concentrations of silicate rocks. The orbit of Pallas makes it unusual among the asteroids because it is highly elliptical with an eccentricity of 0.231 compared to Earth's almost circular orbit and eccentricity of 0.02 ('eccentricity' refers to the roundness of a circle: a perfect circle has an eccentricity of 0 and the number increases as the ellipticity increases). Most of the asteroids in the main belt have broadly circular orbits too, but the orbit of Pallas takes it from a perihelion distance of 315 million kilometres all the way out to an aphelion distance of 510 million kilometres. When these figures are compared to the main bulk of asteroids within the belt then Pallas swings from the inner boundary of the belt, which is around 309 million kilometres from the Sun, to a point about 20 million kilometres beyond the outer limits. Not only does it have a highly elliptical orbit, but its orbit is more inclined to the plane of the Solar System than any other asteroid. Generally, the maximum orbital inclination of individual asteroids is no more than about 30 degrees, but Pallas follows a path which is tilted by a little over 34 degrees. Another unique property of Pallas is the angle of tilt of its axis of rotation, which is estimated to be around 70 degrees. This means the asteroid is almost rolling around the Solar System in much the same way that the giant planet Uranus does.

Like the vast majority of asteroids, Pallas has an irregular shape and has not achieved hydrostatic equilibrium, which means it is resigned to the classification of an asteroid rather than a dwarf planet. With millions of these asteroids in orbit around the Sun you would think that flying through the belt is a hazardous activity. Certainly if science fiction films are to be believed you are currently sweating at the *Kaldi*'s controls, trying to navigate your way through, swerving and dodging around to avoid impact after impact. Thankfully, the reality of travelling through the asteroid belt is a little less exciting than these Hollywood rollercoaster rides. Traversing the 180-million-kilometre-deep belt is still a pretty nerve-racking experience, and it is a journey that takes just over a month to complete, but even though there are millions of rocks out there, they are spread over an area spanning around a trillion square kilometres so the chances of being hit by anything significant enough to cause a catastrophic failure are slim. In fact it's unlikely that you'll even see any of them directly from the *Kaldi*. The larger pieces are well studied and their orbits are well known, but there is always the likelihood that there are other big pieces that have simply not yet been discovered, and many of the asteroids are really quite dark so spotting them against the blackness of the sky is difficult. Furthermore, any object on a collision course with you will have a fixed relative position to the *Kaldi* so will remain stationary in your field of view, making it even harder to spot – it will just get bigger and bigger as it moves closer.

And it's worth remembering that even the smallest items can cause damage. The space shuttle had to have its windows replaced on more than one occasion after impacts that cracked the outer layers. On investigation it was revealed the damage was caused by nothing more than flecks of paint travelling at speeds in excess of 28,000 kilometres per hour. At those speeds even the smallest objects are a serious danger. Things are extremely cluttered around Earth's orbit so a lot of effort has gone into developing protection strategies. For instance, the International Space Station uses a multi-layered skin as protection from the half a million or so pieces of debris about the size of a marble that are flying about. The outer layer is an aluminium alloy and is the first layer of protection, but anything that gets through that would then hit a thick woven fabric much like Kevlar which would absorb most of the energy, slowing the speed of the impactor to a low enough velocity to prevent it from penetrating the interior aluminium skin.

This multi-layering of the shell of spacecraft is common nowadays, but larger obstacles require special treatment, and that largely comes down to getting out of their way. Using radar technology it is possible to track accurately the position of the larger, more damaging pieces of rock and ultimately make the necessary changes to the trajectory of the spacecraft to avoid collisions. And rest assured, the *Kaldi*'s radar will be running 24/7 to pick up any moving objects long before they pose a threat to you.

SIX

A Goliath Among Planets

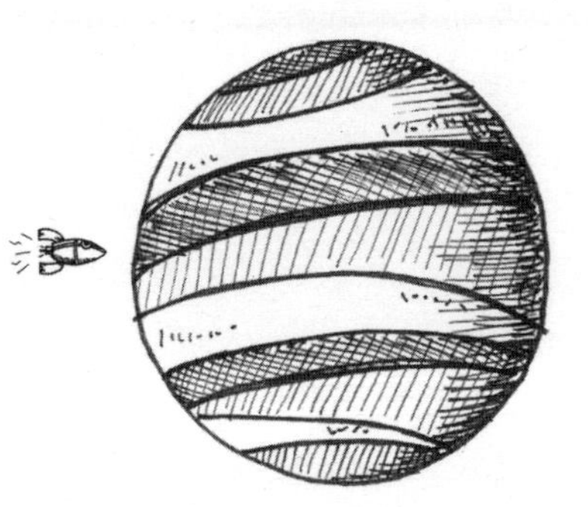

LEAVING THE ASTEROID BELT behind marks a new phase in your journey. With the exception of the Sun, all the objects you have visited so far have been rocky bodies with a solid crust on which you could move around, and they all belong to the group known as the inner planets. Beyond the asteroid belt is the realm of the gas giants Jupiter, Saturn, Uranus and Neptune, and it is to these planets our attention must now turn.

It was the invention of the telescope that really helped us to understand the secrets of the gas giants. Jupiter, the next destination, was particularly interesting, and not only did telescopic study reveal a little about the nature of its atmosphere and the many moons in orbit around it, it also led to the discovery in 1676 of the speed of light.

The discovery of the speed of light goes back to the hunt for a solution to identify longitude while at sea. To solve that problem, Galileo proposed the idea of taking accurate timings of the eclipses of the moons of Jupiter, which he had discovered earlier in the seventeenth century. The idea stemmed from the fact that all these moons orbit around Jupiter in a very regular fashion and, given that the key to pinpointing your location on Earth relied on accurately telling the time from your location, eclipses of the moons would be a more reliable clock than the man-made clocks of the day. Accurate tables were produced that predicted the numerous eclipses over the years and, in principle at least, by observing an eclipse you could work out the time. A great idea, but a flawed one given that accurate telescopic observation from the deck of a heaving ship was difficult if not impossible.

Several decades later a young Danish astronomer by the name of Ole Rømer was working with colleagues to accurately time some of the eclipses, and in doing so he made a rather startling discovery. He made a series of observations between March 1672 and April 1673 which revealed that the eclipses were taking place later than predicted. By estimating the position of Jupiter and Earth in their orbits as he recorded the timings he was able to deduce that it took light about twenty-two minutes to traverse the diameter of the Earth's orbit. With more accurate observations in the years that followed, better estimates of the speed of light were achieved. In 1809, Jean Baptiste Delambre calculated that it takes light eight minutes and twelve seconds to travel the distance between the Earth and the Sun, leading to a value for the speed of light of just over 300,000 kilometres per second – quite an impressive calculation given today's value of 299,910 kilometres per second. It is incredible that the study of a planet that comes no closer than

630 million kilometres to Earth allowed us to deduce the speed of light.

By knowing the parameters of the Earth's orbit we can time how long it takes Jupiter to complete one orbit of the Sun, and from that we can determine the size of its orbit. Like all planets, the orbit of Jupiter is elliptical, but it has an average distance from the Sun of 778 million kilometres which means it takes sunlight nearly forty-five minutes to get there. Because we are a little nearer at closest approach it takes light from Jupiter just over thirty-seven minutes to get to us. This means that from the surface of the Earth we are looking at Jupiter as it was thirty-seven minutes earlier, so we are essentially looking back in time.

Jupiter is the largest of the planets and you could fit 1,321 Earths inside it, but despite its colossal size the fact that it is a big ball of gas means its gravity at the visible surface is just over twice the pull of gravity on Earth. That may seem pretty weak given its size but its pull is powerful enough for any asteroid or comet wandering towards the inner Solar System to struggle to get past. A great example of the way in which Jupiter is the bodyguard of the inner Solar System was seen in July 1994 when Comet Shoemaker-Levy 9 completed its final dive into the Jovian atmosphere. A look at the data of the comet as it moved revealed that it was actually in orbit around Jupiter and was probably captured from an orbit around the Sun during the later decades of the 1900s. On its last orbit around the Sun it would have wandered just a little too close to Jupiter and was taken, and there it sat for a couple of decades in a highly eccentric orbit that took it even closer to the great planet. With every pass Jupiter's gravity tugged at it, increasing tidal forces with every pull. Eventually one of its closest approaches, in July 1992, took it to within 40,000 kilometres of the tops of the planet's clouds and it was here that the comet got ripped apart by the tidal forces into twenty-one fragments. These fragments were then seen to plunge into the atmosphere of Jupiter over six days in July 1994. The fate of Shoemaker-Levy 9 led to Jupiter being dubbed a 'cosmic hoover'. Its sheer presence attracts many asteroids and comets towards it, keeping us a little safer. Estimates suggest that Jupiter experiences around 5,000 more impacts per year than Earth.

With Jupiter attracting all those lumps of rocks it is no surprise that it is orbited by sixty-seven confirmed moons. Galileo discovered four of them when he turned his telescope to the giant planet over 400 years ago, but since the development of higher-quality telescopes, not to mention the advent of space exploration, many more have been found. The largest of its moons, Io, Europa, Ganymede and Callisto, account for 99.997% of the mass of all the moons put together, so you can see there is a significant difference in their sizes. Europa is the smallest of the four with a diameter of 3,100 kilometres but that dwarfs the next largest, which is known as Amalthea and has an approximate diameter of just 168 kilometres.

The satellites of Jupiter fall into three main groups: the four largest are collectively referred to as the Galilean moons; Metis, Adrastea, Amalthea and Thebe comprise the inner group; and finally there are the irregular satellites. This latter group gets its name from the shape of the constituent bodies, and all are significantly smaller than the rest with a highly eccentric orbit. It is believed that this group, which totals fifty-nine satellites, are all likely to be captured asteroids that have strayed a little too close to Jupiter, just like Shoemaker-Levy 9. Many of the irregular satellites seem to have similar orbital characteristics with an almost identical orbital period, eccentricity and inclination, so it is quite likely that these asteroids were once one object that was destroyed in a collision, the resultant debris scattered around the orbit.

The inner satellites, by contrast, have nearly circular orbits and, as their name suggests, they orbit closer to the planet than their more eccentric cousins. With the exception of Amalthea, the members of this group are likely to have formed out of a vast rotating disc of material that condensed out as the planet formed; Amalthea is thought to be a captured asteroid like the irregular satellites.

The much larger and more regularly shaped Galilean satellites can be seen easily from Earth, and of course from the *Kaldi* as you approach the planet. As you coast through the Jovian system and gaze every now and then out of the window you will start to pick out more and more of the smaller and less obvious satellites, too. They all appear as tiny little discs of light with subtly different colours and hues, yet somehow these tiny alien worlds give you a sense of security: you are now among the realm of the gas giants and there are still places where you could set foot on solid ground.

Like the inner satellites, the Galileans formed out of the disc of material that surrounded the planet in its early stages of formation. Unlike all of the other satellites, though, their large sizes dominate the Jovian satellite system. Of the four of them Io is the closest to Jupiter, orbiting at a distance of just 422,000 kilometres, while Callisto is the most distant, 1.8 million kilometres away. At those distances it is thought that they would still have been orbiting within Jupiter's dust disc in the early stages of the system's formation. One theory has it that there have actually been a number of generations of Galilean satellites with each one destroyed through the drag exerted on them by the material in the disc. As each generation was destroyed, another was slowly formed from the debris until such a time that the material in the disc dissipated enough that it no longer affected the moons.

The Galilean moons are physically quite different from each other. Io, for example, is home to nearly 100 mountains (some of which are taller than Mount Everest) and an estimated 400 active volcanoes, making it the most geologically active object in the Solar System. The volcanism is caused by tidal forces from Jupiter and the other Galilean satellites constantly pulling on Io from different directions, leading to tidal heating in the interior of the moon. Io completes one orbit of Jupiter in just over forty-two hours; for every two orbits it completes, the next moon Europa completes one orbit, and for every four orbits of Io, Ganymede completes one orbit. These 1:2:4 orbital resonances are one of the key driving forces for the tidal heating and a tidal bulge that measures around 100 kilometres at its maximum. The volcanoes of Io often send plumes of sulphur and sulphur dioxide high into the rarefied atmosphere which then settle back down on the surface, where they can be seen as dark black streaks. The resultant lava flows and deposits are responsible for the colourful appearance of the moon: the sulphur-based compounds scattered over the surface in red, yellow, black and even green make it look like a cosmic pizza.

Quite unusually for a moon in the outer reaches of the Solar System, which are usually high in silicates and water ice, Io is composed of rocky silicates and iron. As with the planets Mercury, Venus and Mars, it has been possible to deduce the internal structure of Io through its gravitational interaction with spacecraft like Voyager and Galileo. The results show a moon that is differentiated with an iron and sulphur core surrounded by a silicate mantle and crust. The presence of a magnetic field suggests that there is a magma ocean under the crust of Io at a depth of around

50 kilometres, which also helps to explain the high volumes of volcanic activity.

In contrast to the fiery world of Io, the smallest of the Galilean moons, Europa, is thought to have a frozen icy crust under which may be an ocean of liquid water. Europa is a moon that at 3,100 kilometres in diameter is a little smaller than our Moon, but it differs quite significantly in composition. As you approach Europa, you'll see its surface reveals a world markedly different from any other moon in the Solar System. It has a smooth, almost marble-like appearance that seems to lack the craters and mountainous detail of other bodies. What will be evident is its highly reflective surface: Europa has an albedo of 0.64, making it one of the more reflective objects in the Solar System (a perfectly black surface reflecting no light has an albedo of 0; a surface reflecting all light that falls upon it has an albedo of 1). You'll also see a series of darker lines criss-crossing over the surface. These features are known as lineae, and high-resolution images show them as cracks in the crust where the surface material on either side of the lines has moved relative to each other. Other bands show brighter central regions, suggesting that newer material has risen to the surface as the cracks widen in a process similar to the one seen on Earth along oceanic ridges. The cracks are thought to be the result of a flexing of the moon as tides move around as the planet rotates. Given that Europa is tidally locked with Jupiter so that just one face always points towards the planet, it is reasonable to expect a fairly regular pattern of cracks. With the more recent cracks this is indeed the case, but with the older, less prominent features there seems to be a difference which can only be explained if the crust is rotating at a different speed to the interior. This theory is supported by the idea of some form of global sub-surface ocean upon which the crust 'floats'.

The surface temperatures on Europa vary from around minus 160 degrees to minus 220 degrees at the equator and poles respectively, and it is these low temperatures that keep the surface frozen solid. Internal tidal heating would melt the sub-surface layers creating an ocean up to 30 kilometres below the thick icy crust and around 100 kilometres deep. Further evidence for such an ocean comes from the few large craters visible on Europa which seem to be surrounded by ripples, as though the heat of impact temporarily caused the surface ice to become partially melted and the crater's base has been filled with relatively fresh ice.

The idea of a sub-surface ocean gives us another tantalizing possibility of alien life. At the bottom of oceans on Earth there are hydrothermal vents out of which heat is escaping from deep within the planet. Examples of these vents can be seen along the Mid-Atlantic Ridge where two tectonic plates are diverging and new oceanic crust is being formed. No sunlight can penetrate to these depths, yet to scientists' surprise there are entire ecosystems whose source of energy comes from the vents instead of the Sun. It is just possible that deep under the surface of Europa similar oceanic vents are home to a whole new form of life. This is of course conjecture – as yet there is no evidence – but it is entirely plausible. Even though Europa is smaller than Earth, it is estimated that there is around twice the amount of water locked up in the ice and ocean. This makes it a great place for a stop-off on a voyage around the Solar System, to top up dwindling water tanks.

The other two Galilean moons, Ganymede and Callisto, are respectively the largest and second largest of them all. Like Io and Europa, Ganymede is part of the Jovian resonance system where the three are locked into the orbital pattern of 1:2:4 so it also experiences internal tidal heating and it is likely that it too has a sub-surface ocean, this time nearly 200 kilometres below the surface. The surface of Ganymede seems to be split into two different types of terrain: a darker region that is peppered with impact craters – suggesting an age of nearly 4 billion years, making it one of the oldest surfaces in the Solar System – and a lighter region covered in lines and grooves, thought to be a little younger. One of the unique properties of Ganymede is that it is the only moon in the Solar System to have a magnetic field, thought to be driven by convection in the liquid core. The field is buried within the much larger magnetic field of Jupiter and is only visible as a mere disturbance.

Callisto is different because it is not part of the 1:2:4 orbital resonances that affect the other three moons. This means that Callisto does not experience any internal tidal heating and has a very different internal structure. The core is thought to be silicate and surrounded by the mantle, which is 50% rock and 50% ice. Owing to the lack of heating and the inactive silicate core it is unlikely that Callisto has a sub-surface ocean. The surface is old, perhaps as old as Ganymede's, and scarred with impact craters. There is also evidence of frost having formed where ice first sublimated before freezing back on the surface. At the distance it orbits Jupiter, 1.8 million kilometres, it does not suffer from as much radiation as the other moons. This makes Callisto a likely place for some kind of future human outpost in the outer Solar System.

Human exploration of the inner Solar System is relatively easy because the distances are considerably smaller so trips out from and back to Earth can happen in a matter of just months. Trips to the outer Solar System are much more difficult, especially in terms of planning a return to the Earth, because of the immense distances and the time it takes for launch opportunities to come back. The setting up of an outpost that could be used for refuelling and resupply would make the possibility much more realistic. Not only are radiation levels low on Callisto, but the moon is geologically stable with high quantities of water reserves. It's the perfect place. There is even the immense gravity of Jupiter to give any departing spacecraft an immediate swift boost of energy from a planetary fly-by straight after launch. NASA is already looking at using Callisto in this way in their project called HOPE (Human Outer Planets Exploration). It really is only a matter of time before a base like this is scheduled for installation.

You are the first space explorer to journey to Callisto, and therefore the first to have the eerie experience of seeing Jupiter sitting motionless in the sky. Callisto is tidally locked with Jupiter and it takes 16.7 days to complete one orbit; but it also takes 16.7 days for Jupiter to rotate once on its axis, so the planet stays in the same part of the sky, hour after hour, day after day and month after month. It would be an amazing sight, with Jupiter appearing nine times as large as the full Moon from Earth. The detail would be incredible . . . and it would be even more impressive from the surface of Io, appearing thirty-eight times larger than the full Moon and covering an area of the sky around 19 degrees.

Passing by all the Galilean satellites, our trajectory takes us on a close fly-past of the giant planet at a distance of around a quarter of a million kilometres, which puts us just outside the outer ring. Saturn is well known for its splendid ring system but Jupiter also has such a system, although it is nowhere near as impressive. The rings around Jupiter differ from Saturn's not only in appearance but also in composition, as they are made up almost exclusively of dust rather than ice.

There are four main components: an inner torus-shaped halo, the main ring, and then two outer rings. The halo is the nearest to Jupiter with an inner boundary at a distance of about 30,000 kilometres from the cloud tops; it then extends for a further 30,000 kilometres where it meets the inner boundary of the main ring. It varies in thickness and has a shape reminiscent of a wedge, with the thin end nearest the planet. The appearance of the halo, which varies depending on the direction it is viewed from, suggests it consists of dust particles that are no more than 0.015 millimetres in diameter, although smaller particles have been found some distance away from the ring plane. By studying the optical depth of the halo, which is simply a measure of its transparency, it is possible to deduce that the particles have the same properties as those from the main ring and are likely to have migrated from there and be slowly drifting towards Jupiter.

The main ring is the brightest and thinnest part of the ring system with an inner boundary at the outer edge of the halo and an outer boundary just 6,500 kilometres further out. This distance coincides broadly with the orbit of the inner satellite Adrastea, which clearly makes it a shepherd moon of the ring. Shepherd moons are key to maintaining the sharply defined structure of ring systems as their gravity acts upon the particles, keeping them in their orbit. Any that wander out too far will be tugged on by the shepherd moon, causing them to slow down and drop back into the ring, while others will be accelerated and ejected from the system. If the lighting conditions are right then it is possible to detect a fainter ringlet just beyond the orbit of Adrastea. Another shepherd moon, Metis, orbits within the confines of the main ring and is responsible for the evolution of a gap just 1,000 kilometres inside the outer boundary. The presence of a shepherd moon within a ring will cause ring particles to be ejected from the path along the moon's orbit.

The appearance of the rings varies with the direction of the light. There are chiefly two ways a ring system can be illuminated, and they are known as back scattering and forward scattering. Back scattering is the reflection of light back in the direction it came from, but it differs from simple reflection because the returning light waves are scattered in different directions rather than obeying the law of reflection, which dictates that the angle of incoming radiation will equal the angle of reflection. This is not the case when light is back scattered. When the rings are illuminated from back scattering, the observer will be roughly between the rings and the source of illumination – in other words, the Sun. Forward scattering occurs when the rings are between the Sun and the observer. Because Jupiter lies further from the Sun than the Earth the rings can only be observed with forward scattered light by visiting spacecraft such as the *Kaldi*. Forward scattering occurs when light is bent or diffracted around particles and is scattered in the direction it was going before it got to the rings.

Particles within the main ring are thought to last for no more than 1,000 years and are either ejected from the system or drift slowly through the halo and into the upper atmosphere of Jupiter. The cause of the slow but certain demise of ring particles is the radiation that is emitted from the planet, resulting in the so-called Poynting-Robertson drag. The effect is also seen in dust particles in orbit around the Sun and can be understood by considering the process from the point of view of the ring particle itself. Due to their forward motion, the emissions from Jupiter seem to come from a position slightly ahead of them. When the radiation is absorbed by the particles there is a net force acting in the opposite direction to their orbital motion. As a result of the drop in speed, the particles very gradually follow a spiralling path into Jupiter, limiting their time in the rings. If this process takes no more than 1,000 years for each ring particle then there must be some replenishment of particles in the system. One possible source of new dust particles is collisions with the various moons in orbit around Jupiter, either moon-on-moon (rare) or meteoroid-on-moon.

Beyond the main ring are the much wider but fainter Gossamer rings which geographically make up the majority of the Jovian ring system. There are two parts, the inner Amalthea Gossamer ring and the outer Thebe Gossamer ring. Both of them get their names from the satellites (Amalthea and Thebe) that orbit at a distance which corresponds broadly with the ring's outer boundary. The particles within the rings come from Amalthea and Thebe, having been ejected by some kind of high-speed meteoroid impact, and like the particles in the main ring they slowly spiral in towards Jupiter as a result of the Poynting-Robertson effect.

The Jovian ring system is a fascinating structure to study but, as we will see when we travel to Saturn, it is far from being the most impressive system of rings in the Solar System. Also, on a journey through the Jovian system it is hard to maintain focus on the moons and rings when the largest planet in the Solar System dominates the view. The sheer size of Jupiter has been breathtakingly obvious for a while now, and its features are so much more prominent when viewed at close proximity.

The concept of a planet being a giant ball of gas seems a little alien. We are all used to experiencing gas as that stuff that cannot be seen, makes up our atmosphere on Earth and is necessary for our very existence. We can even travel through it seemingly unimpeded. Its appearance in the outer Solar System in the form of a vast spherical ball that looks far from invisible is something that takes a little getting used to. It might be reasonable to think that a spacecraft should be able to fly straight through these so-called gas giants, but the reality is very different.

As with many large-scale phenomena in the Universe, we can look to the force of gravity for the explanation of why it is not possible to fly straight through a gas giant. Like all other normal matter in the Universe, gas molecules are attracted to each other by gravity. When the Solar System formed, the majority of the gas in the protoplanetary disc that had formed around the Sun was forced to the outer reaches. Over millions of years, this gas coalesced into local concentrations which became the outer planets. As time progressed, these concentrations attracted more gas, and with their growth the strength of the gravitational field increased, compressing the gas even more. The force of gravity acts from the centre so it forces the gas to make the most geometrically efficient shape possible – a sphere. If you were to adjust course and attempt to fly straight through Jupiter, ultimately you would fail because of the increasing pressure as you descended through its atmosphere.

Think about your experiences on Earth. Standing on the surface means experiencing pressure from the atmosphere pushing down on you, and in real terms this equates to a pressure exerted on your body of 14.7 pounds per square inch. If you were to descend to the deepest part of the Mariana Trench in the Pacific Ocean then this pressure would increase to around 15,750 pounds per square inch – just over 1,000 times more. You would be crushed. Jupiter is about eleven times the diameter of Earth, so even though gas is less dense than water, descending deep into its atmosphere is clearly going to result in extremely high pressures, perhaps even as high as 500 million pounds per square inch. Gas that is subjected to pressures like that will have a high temperature too, estimated at around 36,000 degrees – hotter even than the surface of the Sun. Gas acts in a very strange way under conditions as extreme as this. As you attempted to fly straight through Jupiter you would first encounter gas in the upper atmosphere which would then turn into liquid with the increasing pressure before becoming solid in the core. For this reason, it is impossible to fly through a gas giant planet. Even without a liquid or solid internal structure, the crushing pressures and roaring heat would mean an end to the mission.

A much more prudent approach is to fly close by the planet and use its gravity to alter the trajectory of your path instead. You will still get a front-seat view of the stunning detail in the upper atmosphere of the planet. Jupiter is often referred to as having an atmosphere, which is a little confusing given that it is a gas planet. The atmosphere is usually considered to start at a point where the atmospheric pressure is about the same as that at the surface of Earth, and we call this 1 bar. This means the atmosphere of Jupiter is about 5,000 kilometres thick. Even this close up, all you can see is the dense ammonia clouds that encompass the planet, embedded in an atmosphere that is broadly the same as the Sun's with similar proportions of hydrogen and helium.

Along with the moons in orbit around Jupiter, Galileo was the first person to observe the belts of the planet and its beautiful hurricane system. The clouds tend to be separated out into different belts and zones that sit at different latitudes and run parallel to the equator. The belts are the darker features, while the zones are found between them and appear lighter in colour. Spectroscopic studies of the light from the belts and zones reveal that the zones are much cooler than the belts, suggesting that they are upwellings of gas, rising higher into the atmosphere and forming ammonia ice crystals before descending in the dark belts. High-speed winds seem to propagate through the belts and zones with maximum speeds reaching over 300 kilometres per hour. It is not known why there are such well-defined belts and zones, although the high-speed jets driving them are certainly the result of solar and internal heating processes. It may be that the Jovian atmosphere where the clouds exist is quite shallow and overlies a more stable lower layer, or alternatively that it is much deeper than expected and they are the visible effects of deep circulation cells in the lower layers of the planet.

The features in the atmosphere are sufficiently stable to have been there for hundreds of years, and astronomers have named them. Around the equator is the imaginatively labelled Equatorial Zone, which extends from 7 degrees south to 7 degrees north, and beyond that lie the North and South Equatorial Belts extending to 18 degrees north and south respectively. These are followed by the North and South Tropical Zones to a latitude of around 50 degrees from the equator, then the less well-defined North and South Temperate Belts and Zones, and finally the Polar Zones. Although the belts and zones are considered to be pretty stable, occasionally one or more of the prominent features has faded from view. The last time this happened was in 2009 when the Southern Equatorial Belt disappeared, only to return in early 2010. The reason for the disappearance of the belts is not fully understood, but it is more likely that during these periods they are merely being obscured rather than having vanished. It may be that high-level cirrus clouds of ammonia crystals form in the air above the belts, masking them from view for several months on end until the clouds dissipate.

It is not just the belts that show some degree of change. There are other atmospheric features that genuinely are more transient, like the numerous vortices of varying shapes and sizes. They are just the same as the vortices we find on Earth and can also be categorized as cyclones or anticyclones, different only due to the direction of their rotation. The cyclones tend to form as small dark patches and are often given the descriptive term 'brown ovals'. Their rotation is in the same direction as that of the planet but their appearance is not restricted to oval shapes: delicate filamentary structures can often be seen in some regions that also show cyclonic motion. But whether oval patches or filaments, they are usually confined to the darker belts. The anticyclones, by contrast, usually appear only in the zones, as white ovals that can last anything from just a few days to over a century. They tend to stay at the latitude within which they formed but do move around the disc of the planet, merging when they meet.

There is one famous anticyclone which appears significantly different to the others, the Great Red Spot (GRS). It was first observed in 1831, having been discovered by the German astronomer Samuel Schwabe, and like all anticyclones on Jupiter it rotates in an anticlockwise direction, taking about six Earth days to complete one revolution. What makes the GRS so impressive is its size: it measures a little over 24,000 kilometres from east to west and about 13,000 kilometres north to south. That may not sound big by astronomical standards but remember, this is a storm, a storm large enough to engulf the Earth . . . twice.

Interestingly, though, it seems to be shrinking. About a hundred years ago it measured nearly 40,000 kilometres east to west – getting on for twice as wide as it appears now – so at that rate the GRS may eventually become circular in shape and might even disappear. However, a conflicting study at the turn of the twentieth century suggested that it is unlikely to disappear completely because of the way it interacts with the surrounding atmospheric features. The study concentrated on the observation of clouds within the storm, and although the spot showed signs of reduction in size, the velocity of the clouds showed no sign of changing. This suggests the storm was as active at the end of the ten-year study as it was at the beginning. It may be that the environment surrounding the GRS has more to do with its shape and size than its evolution. Quite how it has survived for nearly two centuries is one of Jupiter's greatest mysteries, but it is likely that it owes its longevity to swallowing up smaller vortex-like disturbances in the Southern Equatorial Belt (SEB) within which it resides, and to energy being fed to it as warmer air is dragged up inside the storm.

Studies of the spot through infrared telescopes have revealed that it is much colder than the majority of other clouds and atmospheric features on the planet, suggesting it stretches to a higher altitude than other visible features in the area, perhaps by as much as 10 kilometres. They also show that a jet stream blowing eastwards sits to its south, and a more powerful westward jet stream blows to the west, confining the GRS to its latitude of around 22 degrees south. Wind speeds around its perimeter vary but have been recorded to peak at just over 600 kilometres per hour, far greater than hurricane force 5 winds on Earth. It may be windy around the outskirts of the GRS but, like similar anticyclones on Earth, there is little or no wind in the centre – the eye of the storm. Observing the storm's eye in far infrared wavelengths has shown that it is hotter than the surrounding atmosphere because it is dragging warmer air up from lower down in the atmosphere.

The GRS varies in appearance quite substantially from pale pink to a deep salmon colour, although why it is this colour is still not known. There have been occasions when it has disappeared from view, the only clue to its presence a chunk taken out of the SEB. The cause may be the presence of organic compounds like sulphur, but certainly it seems to be somehow related to temperature. The centre of the spot is generally a deeper red than the surrounding regions which ties in with the temperature affecting it in some way. What is known is that it is related to the visibility of the SEB: when the SEB is bright or even white then the GRS is at its darkest, but when the SEB is darker, the spot tends to go light or even disappear.

There is another smaller storm that is known as Oval BA – or, more affectionately, Red Spot Jr. It is also found in the southern hemisphere but a little further south in the Southern Temperate Belt. Wind speeds around this storm have reached 618 kilometres per hour, which is comparable to winds around the Great Red Spot. Back in April 1996, Tropical Cyclone Olivia battered Barrow Island in Australia with winds gusting to record speeds of 407 kilometres per hour. Compare that, which was a wind gust, to the sustained wind speeds around the storms of Jupiter and you'll get an idea of just how violent they are.

When you are travelling on a plane at home, the pilots will do anything they can to avoid flying through giant storms as the experience for you, the passenger, would be pretty unpleasant. Larger aircraft are clearly more stable, but even for you in your spacecraft the GRS presents a serious challenge. If you tried to fly through it at a decent altitude then the first thing you would experience on the approach would be turbulence, starting off as small lumps and bumps but getting more severe as the flight continued. As you got closer to its outer wall the tail winds would pick up to speeds around 600 kilometres per hour with a strong component coming from the left, blowing you to the right. Anyone observing from outside the storm would see you appear to accelerate rapidly at this point as the wind carried you along, and at this stage it would be pretty impossible to abort as the wind speeds would seriously inhibit any means of escape.

As you pressed on into the GRS you would be hit by very strong downdrafts, forcing you down at high speed. This is one danger that aircraft face when flying through storms close to the ground. It's fine at altitude as there is plenty of time to recover, but experience a downdraft close to the ground and the likely conclusion would be 'early contact with the ground' – in other words, a crash.

Assuming you could regain control from the downdrafts, windshear would be the next and perhaps most dangerous challenge. Windshear is the variation of wind either vertically or horizontally, and the greater the variation the stronger the windshear. Around the outside of the GRS are the downdrafts just encountered, but dragging warmer air from below are updrafts that pierce through the central column of the storm. Flying from violent downdrafts into equally strong and possibly worse updrafts is the most perilous sector of the flight. At this point there is even a risk of structural failure. The *Kaldi* may well get ripped to shreds – which is a very good reason for keeping this journey hypothetical.

Pass instead at a safe distance, thousands of kilometres above the GRS, and you will notice that there is a serene beauty about it. Below the almost mesmeric cloud belts and features in the atmosphere of Jupiter is an alien world unlike anything you have encountered so far. Descending through the Jovian atmosphere, there are four distinct regions which bear identical names to the regions in the atmosphere of the Earth: the exosphere, the thermosphere, the stratosphere and finally the troposphere. Like all planetary atmospheres there is no sharply defined boundary where the atmosphere finishes and space begins; instead there is a very gradual transition from the density of gas in the atmosphere and the gas-deficient but not -devoid realms of interplanetary space.

Observations have been made of aurora activity and airglow in the thermosphere of Jupiter. Aurora is a familiar phenomenon – you saw it in the atmosphere of Earth just after you left – but airglow is a new concept, although it too is sometimes visible over the Earth. It is caused by incoming sunlight stripping electrons from molecules in the atmosphere, and then, as they try to reattach to the atomic nuclei, they give off a tiny amount of light and cause the gas to glow.

As the descent into the atmosphere continues, the temperature profile is not what might be expected. Instead of a gentle increase, the opposite is true: it gets cooler the lower you go. This is because the rarefied gas at the top of the atmosphere easily absorbs solar radiation, taking temperatures as high as 1,000 degrees. If you were to hop outside the spacecraft it would still feel cold though because the space between the atoms is so high that heat transfer is non-existent.

At an altitude of about 320 kilometres the top of the stratosphere is encountered, and at this point the temperature decrease stops and a fairly constant 100 degrees is maintained all the way to the tropopause at an altitude of about 50 kilometres. The troposphere, through which the descent now continues, is the region where the majority of the belts, clouds and storms are seen, and now the temperature slowly starts to increase again, to around 400 degrees at the 'surface'.

The Jovian clouds that can be seen from Earth are probably the most complex seen anywhere in the Solar System. In the upper regions of the troposphere they are made mostly from ammonia ice and ammonium sulphide. Below these are clouds of water, and it is their presence that has had a big influence on the atmospheric conditions. It takes more energy for a given amount of atmospheric pressure to transform water from vapour into a gas compared to ammonia, which, along with the higher quantities of water, causes huge amounts of energy to be transferred to and from the atmosphere.

At the bottom of the troposphere, the pressures and temperatures are so extreme that they are above the point where hydrogen and helium blend seamlessly from a gas to a liquid without a solid phase in between. They are said to be supercritical fluids in this state as they just get denser and hotter the lower you go. This is thought to continue all the way through to the core of Jupiter, where, as we know, it may become so dense as to be a solid metallic core. It is also likely that convection currents within the supercritical liquid layers may have mixed with a metallic hydrogen core, redistributing it within the interior of the planet. For now, the exact nature of the interior of Jupiter remains a little bit of a mystery.

Some interesting stories have been written about the Jovian atmosphere that suggest life may have evolved among the clouds. Any creatures could only survive high up in the atmosphere because of the higher pressures lower down. That means they would need to float constantly, so the idea seems a little far-fetched. Even if the organisms were able somehow just to float they would need to be resistant to extreme levels of solar radiation. And all this, of course, assumes a nice, stable, almost quiescent atmosphere that allows them to float, as if on a gentle breeze. As we have already seen, the conditions in the Jovian atmosphere are very different; jet streams, vortices and violent vertical winds could easily suck any life form into the crushing pressures down below. Even if they could survive the extreme pressures, the temperatures would make the environment completely inhospitable.

There is, however, one hardy little micro-organism that gives some hope for finding life elsewhere in the Solar System. It is known as a tardigrade and measures just half a millimetre when fully grown. Research has shown that these tough little critters can survive pressures of around 150,000 pounds per square inch, which is about ten times the pressure found at the bottom of the Mariana Trench so they would certainly be capable of surviving at lower depths in the Jovian atmosphere, but not strong enough to survive the million-pounds-per-square-inch pressures down in the liquid metallic hydrogen levels. They can withstand temperatures in excess of a few hundred degrees, and also significantly more radiation than us lowly humans. They can survive in the vacuum of space and are able to do so without any food or water for about ten years. These creatures are *tough*. But still it seems even they cannot survive in the atmosphere of Jupiter. It looks like the only place where life may be able to evolve around here is in the sub-surface oceans on some of the moons.

As you swing past the planet on the closest approach you get a boost, trading some of Jupiter's orbital velocity for spacecraft speed. It is amazing to think that the *Kaldi* is not the first spacecraft to be here and to do this. Pioneer 10 was the first of a flotilla that have visited the outer Solar System, having been followed by Pioneer 11 and Voyagers 1 and 2. Pioneer 10 was launched from Cape Canaveral in Florida on 3 March 1972. In February the following year it became the first spacecraft to fly through the asteroid belt, which it did suffering no damage at all. It first encountered the magnetic field of Jupiter on 16 November 1973, when it made the startling discovery that the magnetic field was actually inverted compared to the Earth's.

Planetary magnetic fields like Jupiter's and Earth's have two poles, a north and a south, just like the bar magnets used in schools. Earth is actually the peculiar one rather than Jupiter because its north pole sits under the geographic south pole while the south pole of the magnetic field is in the northern hemisphere. That is why the needle marked north on a compass points north, because it is attracted to the south magnetic pole. When the fields reverse, the north needle will point in the other direction and all compasses will have to be remade. Jupiter has its magnetic poles under the geographic poles where you would expect them, but it is not unusual for planetary magnetic fields to reverse. It is even possible to study past magnetic reversals by studying the magnetic properties of metallic rocks. There have been a number of reversal events on Earth over its history. The last major one happened 780,000 years ago when the field strength dipped by just over 5%.

Having been in the neighbourhood of the mighty Jupiter for a total of sixty days and having studied the magnetic field of the planet and made numerous discoveries about new moons and atmospheric features, Pioneer 10 finally went on its way. Unlike yours, the path it took then sent it off and out of the Solar System, away from the other outer planets. At the speed it is flying at – around 12 kilometres per second (most commercial jets travel at about 0.2 kilometres per second) – it could travel the diameter of the Earth in just over seventeen and a half minutes. That sounds pretty impressive, but the distances between the stars where Pioneer 10 is heading are mind-bogglingly vast. It is now flying towards the bright red star in Taurus known as Aldebaran, sixty-five light years away (which of course means that when we observe Aldebaran from Earth, we are seeing it as it was sixty-five years ago). Travelling at its current speed it will take 2 million years to get there.

For you, though, the outer planets are very much on the agenda. Your encounter with Jupiter is over and it is on with the journey and another long haul of empty interplanetary space for around 500 million kilometres. At your current speed, that will take a little over a year. Clearly interplanetary space travel is not for the faint-hearted, but to be able to see the beautiful and enigmatic Saturn, the jewel of the Solar System, at first hand is going to make it all worthwhile.

SEVEN

The Jewel of the Solar System

IT HAS NOW BEEN just over four years since you left home and you've no doubt thought many times about all those comforts you left behind: home-cooked meals, a decent hot shower, a nice pot of tea or pint of beer, fresh air. Mealtimes are a vitally important part of any space mission as they are a familiar routine that has both physical and psychological benefits. Of course we all recognize that we need to eat and drink in order to function efficiently, and to that end there are two important elements: nutritional value and calorific value. The latter is the energy you get from food and drink so sufficient calories are required for that purpose, but if there are insufficient nutrients in meals then your ability to focus on tasks will be affected, as will your general performance and, ultimately on a long voyage like this, your health.

Having food that provides all the goodness required by the human body is one thing, but the food must also be acceptable if morale is to be kept high. The acceptability of food is based on the experience of eating, so it's all about the way it interacts with our senses. Taste, texture, smell and appearance are of almost equal importance to its nutritional and calorific value.

So, having regular good-quality meals is essential for space travellers such as yourself, even though in the confines of a spacecraft this is no mean feat. Pre-packaged food has been used almost exclusively on board the International Space Station, and with a shelf life of a year and a half it serves the purpose well. Unfortunately a huge amount of waste is produced this way: many of the space shuttle missions reported that 80 to 90% of their waste products was food packaging. When there are regular visits from supply ships this is not a problem as they can transport waste back to Earth for responsible disposal, but on a long space journey such as yours, waste must be kept to a minimum. In fact, your journey around the Solar System will take nearly fifty years to complete so it has not been possible to load on board enough pre-packaged and dried food to sustain you. Not only would it be unlikely to have a long enough shelf life, that amount of food and its waste products in storage would add far too much mass to the *Kaldi*. The Apollo missions to the Moon in the 1970s allowed 1.1 kilograms of food per person per day. In order to keep just one person fed on a fifty-year voyage would therefore require 20 tonnes of pre-packaged food – about as much as the weight of ten average family cars.

Clearly long-term space exploration can only happen with some type of system to facilitate the production and growth of food. There has been extensive research into systems like these, and by far the most promising is known as hydroponics. This method relies on growing plants in a liquid nutrient solution without the use of soil. It has many benefits, the greatest among them that watering is not required as the water stays within the system, and the nutrient levels can be completely controlled within the solution allowing for far more efficient cultivation. There is a fully functioning computerized hydroponics nursery on board the *Kaldi* which has been providing you since launch with a range of fruit, vegetables and herbs, and it'll continue to do so for the duration of the journey. The only way to include meat in the diet would be to keep some sort of farm on board, which for obvious reasons is wildly impractical. For the *Kaldi*'s trip, then, we have settled for a vegetarian diet with food supplements to provide whatever additional nutrients you require.

This doesn't quite satisfy the requirement mentioned earlier – about the importance of taste, texture, smell and appearance – but don't forget, we chose you to go on this mission because you were made of hardy stuff. You're certainly able to prepare and cook it all in much the same way as you did back on Earth. Because of the simulated gravity environment, conventional appliances such as a microwave can be used in a conventional way – though electric or gas ovens have been avoided because they increase the risk of fire, and one thing you do not want in a spacecraft is a fire. There are of course smoke detectors on board, but best they don't go off in the first place.

Water flows freely out of taps, too, though to provide enough water for a single person for just fifty years would mean carrying 220 tonnes of the stuff in tanks – clearly not an efficient way to remain hydrated. Fortunately there is a water recycling system in place. Living creatures, anything that lives and breathes, drink water and then recycle it by exhaling it, sweating and urinating. All this water can be collected, cleaned and purified by a recycling system like the one on the *Kaldi*, so you have always had fresh drinking water. Not a drop is wasted. There should be no need to top things up, but if this was necessary there are plenty of places in the Solar System where water supplies can be replenished – the polar caps on Mars, the deep craters of the Moon, inside the gas giants, the many icy moons of the outer planets, and even comets. Having sufficient water on board the *Kaldi* has of course allowed you to indulge daily in one of your luxury items – coffee.

As you sip your latest cup and munch on your most recent culinary creation – a carrot, fennel and broccoli stir-fry – you stare out for the umpteenth time at what has been for a long while now a largely uninspiring view. It's easy to understand why the good old *Starship Enterprise* had a holodeck on board to give its occupants some means of escaping the monotony of long voyages. Unfortunately holodeck technology is a little too far-fetched for us, but thanks to the increasing capacity of digital storage there is a vast library of your favourite television shows and movies for you to choose from and watch on the huge plasma television with surround sound. There is plenty of music to listen to as well, and the folks on Earth have been sending regular audio news bulletins to keep you up to date on what is going on back on your home planet.

You have been dipping into this library a little more often of late because this section from Jupiter to Saturn has been the longest leg of the journey so far – but there are longer ones ahead. The 'sky' is looking blacker than the blackest nights you will remember on

Earth, the stars are looking brighter than ever before, and Jupiter is getting ever smaller . . . but being slowly replaced by a growing Saturn. As you approach the planet, it starts to appear slightly oval in shape, but essentially it is the view you may have seen through a telescope from Earth. The rings are finally visible in their full glory, but you will not be able to make out any moons just yet.

Saturn is the second largest planet in the Solar System with a diameter of 120,000 kilometres across the equator and 108,000 kilometres across the poles. It rotates on its axis around once every ten hours and thirty-four minutes (this is about thirty-nine minutes slower than Jupiter, which rotates once every nine hours and fifty-five minutes on average), but that very much depends on the latitude of features which rotate faster around the poles than at the equator. The fast rotation means it bulges out at the equator, taking on the appearance of a squashed ball – an appearance shared by most of the other planets but to a lesser degree than Saturn. There is a wonderful fact about Saturn: regardless of its monstrous size, its average density is low enough that if you could find a body of water large enough, Saturn would float. Although it is thought that there is a rocky core, it is the extensive atmosphere that reduces its average density to 30% lower than that of water.

On arrival at the sixth planet from the centre of our Solar System, the Sun appears only as a very bright star. It now lies 1.4 billion kilometres away, which is nine and a half times further than the Earth is from the Sun. It takes sunlight seventy-nine minutes to reach Saturn and seventy-one minutes for that light to reflect back from Saturn to Earth. One of the things you will have noticed is the increasing difficulty of communicating with Earth. Radio waves propagate through space at the speed of light so any message you now send to Earth, even just saying 'hello' through an interplanetary telephone, takes seventy-one minutes to get there; assuming an immediate response, it'll be the best part of two and a half hours before you hear a 'hello' coming back. The further away you travel, the fainter the signal is getting too, and that will be making feelings of isolation grow stronger every day.

But it is easy momentarily to forget such things when you're close to a planet like Saturn. Being the first human to see it like this will be an amazing and emotional experience. It's easy to imagine what the Apollo 8 astronauts must have felt when they became the first people to leave Earth orbit and fly around the Moon. To be the first to visit this alien world, to be the first pair of human eyes to gaze upon Saturn and its beautiful ring system . . . words alone will not be able to describe the feeling.

From seeing Saturn through a telescope from Earth you may recall how large the ring system looks, and also that beyond their boundaries was a tiny speck of light – the planet's largest moon, Titan. There are now sixty-two officially identified moons in orbit around Saturn of varying sizes, some measuring just a few kilometres across. Titan, however, is well named: it is larger than the planet Mercury. They are all divided into ten groups based on their orbital properties, from ring shepherds to co-orbitals and ring moonlets to large outer moons. As their name suggests, ring moonlets and ring shepherds orbit in the vicinity of the planet's ring system. The moonlets differ from the shepherds owing to their impact on the rings, the latter creating very well-defined gaps in the rings – the Encke Gap is one of them – while the moonlets create only partial gaps. Co-orbitals share broadly the same orbit so that they gravitationally interact with each other. There are actually only two co-orbital moons, the larger Janus and the slightly smaller Epimetheus, and their orbits are 151,460 kilometres and 151,410 kilometres from Saturn respectively. Another of the groups is imaginatively known as the inner large moons and includes Mimas, Enceladus, Tethys and Dione, all of which orbit in the so-called E Ring of Saturn (there's more on these rings later). Orbiting between Mimas and Enceladus are three moons that fall into a different group known as the Alkyonides. There are then four other groups that chiefly cover small, irregular-shaped moons.

The remaining two groups have some interesting properties. The Trojan group of moons are unique to Saturn and are special because of the orbital relationship they share with the planet and two other moons. There are four Trojan moons: Telesto and Calypso form a system with the inner large moon Tethys, while Helene and Polydeuces are bound to Dione. It is the gravitational relationship with their parent moon which sets them apart from the other small satellites of Saturn. In any orbital system where one object orbits another there are five places where a third object, which is affected only by the force of gravity, can maintain its relative position with respect to the other two objects. These places are known as the Lagrangian points, after the French mathematician Joseph Louis Lagrange who in the eighteenth century made his reputation in the field of celestial mechanics. Three of the Lagrangian points are found on a line between the two objects: L1 is directly between the two objects, L2 is along the line but beyond the smaller of the two, while L3 is along the same line but behind the more massive of them. The final two points, L4 and L5, are at a position where they make up the third point in an equilateral triangle with the line between the other two objects as its base. L4 is the point that lies ahead of the moon and L5 is at a point behind the moon. Saturn's Trojan moons sit at L4 and L5 of the two larger moons.

Just like any other body in the Solar System, the Earth's system has Lagrangian points too and they have been used over the years with some success for certain space missions. The Lagrangian point between the Earth and the Sun at position L1 is ideal for space observatories for solar study because from this point the Sun is never obscured by the Earth or Moon whereas a conventional satellite in Earth orbit would suffer periods of time when the Sun was out of view. The L2 position is a great place to put a more general space telescope, as from here the Sun, Moon and Earth are all relatively closely placed, leaving a huge area of permanently unobscured sky.

The most famous of Saturn's moons are found within the group known as the outer large satellites. The smallest of them is Hyperion with an average diameter of just 270 kilometres – small enough to be able to sit it on the United Kingdom and have room to walk around its perimeter. It is the largest irregular-shaped object in the Solar System and resembles a badly misshapen rugby ball; but perhaps its most intriguing feature is its sponge-like appearance, which comes from the high quantity of deep, high-walled and jagged-edged craters that cover the surface. The largest of these craters is 121 kilometres in diameter and an incredible 10.2 kilometres deep – and bear in mind that the entire moon is just 270 kilometres in diameter. At the bottom of the majority of the craters is a dark, reddish-coloured sediment that is thought to contain carbon and hydrogen compounds – a combination commonly found on other Saturnian moons. Taking the appearance of the moon into consideration, it is very likely that Hyperion was once part of a larger satellite which was broken apart by an impact event. The satellite we now see has a low density which suggests it is likely to be composed mostly of water ice with only small amounts of rock and takes on the structure of a pile of rubble rather than a solid object.

Not only is the shape and appearance of Hyperion a little unusual, so too is its rotation. Unlike all other naturally occurring planetary satellites, it is not tidally locked to Saturn. This means it does not always present the same face to the planet, as does our own Moon. Instead, Hyperion rotates in a rather chaotic and almost random fashion so that its axis of rotation rarely points in the same direction for long. In contrast, Earth's axis of rotation does point to different positions in space but it moves so slowly that it seems to be stably pointing in the same direction for thousands of years.

Iapetus is over five times larger than Hyperion with a diameter of around 1,468 kilometres. Unlike its smaller cousin, it has a much more uniform shape and structure. One of the most impressive features of Iapetus, which orbits Saturn at a distance of 3.5 million kilometres, is its rather strange two-tone colouration. This discovery was made back in the seventeenth century by Giovanni Cassini who noticed that he could see the moon when it was on the western side of the planet but never when it was on the eastern side. This rather curious observation led Cassini to the correct conclusion that it must be tidally locked to Saturn and have one hemisphere that was darker than the other, so when that hemisphere was presented to Earth it became difficult if not impossible to see visually. 'Magnitude' is a term astronomers use to describe brightness as they see it, represented by a number on a logarithmic scale, brighter objects having a negative number. The dark hemisphere, which is the leading face of Iapetus, has an apparent magnitude in the sky of 11.9 but the trailing edge is brighter with an apparent magnitude of 10.2. It just happens that the best telescopes of the time could detect the moon when it was shining at 10.2 but not when it presented its darker face.

The darker region was named Cassini Regio after its discoverer and the lighter area was separated into two areas: to the north of the equator is Roncevaux and to the south is Saragossa Terra. The darker region, which has a reddish-brown hue to it, is thought to have originally looked the same as the lighter regions, which are rich in ice, but the low pressures and temperatures at the surface will have allowed the ice to sublimate – a process we have considered already, where a solid turns straight into a gas but in doing so it leaves behind the rock and dust it was once mixed with. It is now believed that the darker regions are 'lag' or deposits from the sublimation process. Observations from spacecraft and from Earth-based telescopes have shown that the darker deposits, which are no more than a few tens of centimetres at their deepest, are carbon-based but also contain significant quantities of the extremely poisonous element hydrogen cyanide.

Interestingly, because the darker material absorbs more energy from the Sun, the darker regions experience higher daytime temperatures, on average 16 degrees warmer than the lighter regions. This higher temperature around the Cassini Regio means that ice sublimation occurs at a much faster rate while ice deposits are more likely in the colder regions, which ultimately leads to the dark regions becoming darker with more lag and the lighter regions becoming lighter with all exposed ice in Cassini Regio eventually disappearing. For this thermal runaway process to start, there must have been some kind of catalyst, and it is likely that this came from debris falling on to Iapetus, probably from nearby Phoebe. The colour of Phoebe closely resembles the brighter regions of Iapetus, although it would only take a very slight difference in reflectivity for the temperature difference to be sufficient to start the process.

Another strange feature of Iapetus makes it somewhat resemble a walnut. Running for almost 1,300 kilometres (over a quarter of the circumference of the moon) is an equatorial ridge that cuts through Cassini Regio and rises up to 20 kilometres higher than the surrounding plain. This makes some of its peaks the tallest mountains in the Solar System, and the presence of cratering along its length suggests it is very old. There are a number of possible explanations for the formation of the ridge but none of them satisfactorily accounts for its appearance, not least the accuracy with which it seems to hug the equator of Iapetus. One possible cause rests with the moon's rotation period, which may have been a lot higher in previous millennia. If this were the case, then Iapetus may have cooled fast enough to retain a more plastic viscosity, allowing the pull of Saturn's gravity to maintain the height of the ridge. Another competing theory suggests that Iapetus may have retained a ring system soon after its formation which slowly accreted on to the surface around the equator. It remains just another of those outstanding mysteries of the outer Solar System.

Rhea is the second largest moon of Saturn with an equatorial diameter of approximately 1,526 kilometres. It is a pretty typical moon of the outer Solar System with a heavily cratered surface and a makeup of about 75% water ice and 25% rock. One unusual feature was discovered back in 2008 by the Cassini spacecraft as it flew past. It detected a change in the flow of electrons that were trapped by the magnetic field of Saturn and noted a higher concentration of dust and debris around Rhea. There is a region around all astronomical bodies known as the Hill Sphere, and within this sphere the gravity of the more massive body is dominant. Saturn will have a Hill Sphere which has shaped its family of moons and the wonderful ring system, but the moons too will have such a region surrounding them. In the case of Rhea, an increase in density of electrons and dust and debris suggests that it too has a tenuous ring system. The important word here is 'suggests' as there has so far been no direct observation of ring system particles, although ultraviolet observations revealed bright flashes around the equator which may have been the result of ring debris crashing into the surface. If Rhea does indeed have a ring system then it will be the first discovery of a system of rings in orbit around a moon.

The largest moon of the Saturnian system we have met already, and Titan is without doubt one of the most intriguing. An approach to Titan will give away its most unique feature – or rather lack of them. Nearly every other moon in the Solar System will reveal craters as you draw closer but Titan appears as a featureless world because of a dense atmosphere. Not only does it have a thick atmosphere which sets it apart from all other moons, it is also the only body other than Earth in the Solar System which shows evidence of bodies of surface liquid. Not a lot was known about Titan until the advent of space exploration and certainly its atmosphere makes remote observation of the surface pretty tricky. Radar can penetrate the atmosphere and has been used with great success by the Cassini probe, but the most we have ever learned about this strange world came from a lander which Cassini deployed. In 2004, the Huygens probe descended through the atmosphere, landed on the surface and gave us our first real glimpse of this almost prehistoric Earth-like environment.

Of all the places you are flying by on your journey around the Solar System, Titan is most definitely one that is worth stopping off to see, to take a walk on its foreign shores. Such an excursion is subject to many of the challenges faced by spacecraft on return to planet Earth – for example, the thick atmosphere would lead to overheating of any spacecraft trying to penetrate it unless the angle of attack was just right. And the approach angle has to be just right, not only to avoid overheating: if it were too shallow then the craft would skip off the atmosphere like a stone skimming across the surface of a pond; too steep and it would just burn up. The atmosphere is nitrogen-rich, much like Earth's, with hydrogen and methane the other main constituents. The atmosphere is denser than our own so by the time you reach the surface and start to walk around, your body will be exerted to 1.45 times the surface pressure you experience on Earth. The temperature at the surface is about minus 180 degrees but this would be a whole lot lower if it were not for the methane creating a greenhouse effect and warming the climate. Taking that low surface temperature and lack of oxygen into consideration, you will most definitely need to wear your space suit.

The surface is lit by an eerie orange glow not too dissimilar to the surface of Mars, although an awful lot darker. On Titan, the Sun is giving you just 1% of the light it provides back on Earth; add that to the dense atmosphere, and the surface is only very gently illuminated, even in daylight. You have landed along the southern coastline of Ligeia Mare, the second largest of the seas on Titan, which is found in the north polar region. You have landed in this particular location owing to data collected by the Cassini spacecraft. Radar was used on Titan just as it was on Venus because of the obscuring effects of the atmosphere, and you will remember that bouncing a signal off the surface and analysing the echo allows us to interpret the terrain below. When Cassini passed over Ligeia Mare it discovered a sea that was roughly 500 kilometres along its longest axis and surprisingly calm. The resolution of the radar technology on board Cassini would have allowed it to detect waves as small as a millimetre in height, but it could not even detect those.

Looking out over the sea now, you can see it is almost as smooth as glass, reflecting the sky beautifully. There is hardly any wind, which goes some way to explaining the absence of waves, but there may be other causes. The sea is mostly methane liquid with a little ethane and other elements but it looks like it may be covered by some other form of liquid that is suppressing the waves in just the same way that oil-spills back on Earth reduce waves to tiny ripples. If only you had a boat that could set sail on this alien sea that has probably sat undisturbed for millennia . . .

The sea's southern coastline is characterized by a rolling landscape that has been sculpted by millions of years of erosion. Look in the other direction, to the south, away from the sea, and you'll see distant hills with what look like dark rivers running down them. These rivers are actually just dark channels which are the result of organic compounds being created high up in the atmosphere of Titan from the interaction of the gases with ultraviolet radiation from the Sun. A methane-based rain then helps to bring these compounds out of the atmosphere and they wash down the mountains to be deposited in the channels and on the plains of the hills. Rain on Titan, concluded the Huygens probe which landed there in 2004, is probably a fairly rare occurrence.

The surface underfoot is covered in a soil-like material that has an almost bouncy feel to it, a little like compressed snow. Step carefully and you can walk across the surface; tread too hard and you'll probably sink down into softer material. It may even have a wetter consistency further down – a little like a crème brûlée, with a crusty surface and a more soggy substance underneath. There seems to be little evidence of impact cratering from where you're standing but that is consistent for a body with a dense atmosphere which causes all incoming chunks of rock to burn up, letting through only the largest pieces. Those that are present are quite young; the rest of the surface is no more than a billion years old. What there does seem to be plenty of evidence of is volcanism.

On the journey in the RSU back up to the *Kaldi*, analysis of the atmosphere reveals the high quantities of methane in it, and when studied over a period of time the level seems to be fairly sustained. Certainly the bodies of methane liquid on the surface will produce a certain amount through evaporation, but one likely explanation is that volcanic eruptions are injecting further amounts of methane into the atmosphere. The presence has also been detected of an element known as argon-40, which indicates the existence of cryovolcanoes, or volcanoes that erupt lava of water and ammonia.

Despite all the evidence in the atmosphere, there seems to be a distinct lack of direct evidence for volcanic activity on the surface. Among those features that have aroused suspicion are some bright spots that appeared in Titan's atmosphere in 2008. These could easily have been explained by meteorological phenomena but they seemed to last far too long to be weather events. A change in brightness was also detected on the surface in a region known as Hotei Arcus in the southern hemisphere, which was attributed to possible lava flow.

Perhaps the most likely candidate for a cryovolcano on Titan was announced by NASA's Cassini team in 2010. The area, known as Sotra Patera, is a chain of three mountains that rise 1.5 kilometres above the surface and they all seem to have a large crater at the top. The chain nature of the mountains and the craters strongly suggests that some sort of geological process led to their formation, and radar detection revealed what looks like frozen lava flows around their bases.

Perhaps one of the most tantalizing things about Titan is its similarity to Earth when it was in a more primitive state. The moon seems to harbour some really quite complex organic compounds – an environment that is said to be 'prebiotic'. As we have seen, though, the surface temperature is far lower than the average temperature on Earth, at least by a couple of hundred degrees. This low temperature and the lack of surface liquid water lead many to rule Titan out as a possible location for finding primitive alien life, but the conditions could perhaps be tolerable for non-water-based life. It may just be that Titan is another moon that has a sub-surface ocean of water which could support life that would not necessarily be dependent on the Sun for its energy. The oceans of methane could support life too: where life on Earth takes on oxygen and produces carbon dioxide, on Titan it may start with hydrogen, which is plentiful, and produce methane.

If even primitive life like this existed, we might find evidence within the atmosphere of the moon. Levels of atmospheric hydrogen would be reduced within the troposphere, along with a reduction in the levels of acetylene which would form part of the biological process. A report published in 2010 by Darrell Strobel of the Johns Hopkins University in the United States announced exactly that. The study showed how the levels of hydrogen and acetylene reduced with altitude in Titan's atmosphere, giving credibility to the possibility of organic activity. I choose these words carefully since this is not evidence of alien life, just a suggestion that there might be some form of organic chemistry taking place on Titan. More research is needed before we get to the bottom of the true nature of the processes.

Whether life has or will ever evolve on Titan is still one for scientific debate, but one thing that is more certain is that the conditions on this Saturnian moon will evolve as the Sun moves along its evolutionary path. In a few billion years' time the Sun will start to swell as it becomes a red giant star, increasing in size so much that life on Earth may become impossible. There is hope for us, though, because by that time Titan may well offer an alternative location for humanity to evacuate to. When the Sun turns into a red giant, the surface temperature on Titan is likely to increase by about 100 degrees, taking its highest temperature to around minus 70 degrees which is almost 20 degrees higher than the lowest temperature ever recorded at the surface of the Earth. At temperatures like this it may just be possible for oceans of water and ammonia to exist on the surface. Looking into the future, then, Titan is one of the best locations for us to keep a close eye on for long-term exploration and maybe even habitation.

If we were ever to set up a human outpost on Titan, or any of the Saturnian moons, its inhabitants would have the most incredible view. To gaze up at night, and even in the daytime, and see Saturn with its beautiful ring system spanning the sky would leave most people awestruck. Without doubt Saturn is famous for its ring system, which is composed almost entirely of water ice particles with a few rocky components made of carbon-based elements. The particles range in size from a millimetre or less up to about 10 metres, yet from a distance they give the appearance of a stunning system of rings encircling the planet. It is very difficult to give dimensions for the system because some of the outer rings are faint and quite diffuse, but it is generally accepted that the main rings extend for about 270,000 kilometres. Taking into account the fainter outer rings then the entire system spans nearly 26 million kilometres.

The average thickness of the rings is no more than about 100 metres so the whole thing looks somewhat like a giant yet very thin celestial disc. Here's one way to visualize just how thin they are:

if the whole ring system were shrunk down to the thickness of a piece of paper and the proportions kept the same, then the diameter would still be 26 kilometres.

The rings were discovered by Galileo in 1610 when he turned one of the first telescopes to the sky, although he did not realize he had discovered them. Now bear in mind that it was a pretty primitive piece of kit he was using, so the views he got were far from the best. He noted that 'Saturn is not alone, but is composed of three, which almost touch one another and never move nor change with respect to one another'. Galileo became confused in 1612, however, because the two 'ears of Saturn', as he sometimes referred to them, vanished from view, only to return again in 1613. What he had observed was a ring plane crossing. Like a child's spinning top, the planet Saturn wobbles on its axis so that every fourteen to fifteen years the rings are presented to us edge-on. As we have seen, they are very thin, so when appearing edge-on from Earth they seem to vanish from view, only to return some months later. In 1655, the Dutch physicist Christiaan Huygens used an improved telescope he had made with a magnifying power of around ×50 to observe Saturn and became the first person to note that it 'was surrounded by a thin flat ring, nowhere touching the planet'.

Observation of the rings with modern Earth-based telescopes reveal that they are not a solid structure. Indeed, if they were, then they would be unstable and soon break apart. Instead they can be seen to be made of hundreds of smaller rings that are broken up by gaps. The main rings are named alphabetically in the order they were discovered. Starting from the planet, the main rings are D, C and B, then there is a gap known as the Cassini Division, followed by the A ring. Within the A ring is another gap known as the Encke Gap, and beyond are the F, G and E rings.

There are two competing theories for the origin of the rings. The least popular relates to the ring particles being leftover material from the formation of the Solar System. The proportions of ice and rock particles within the rings suggest a different origin. A more popular theory has it that the ring particles are actually some of the remnants of a moon that wandered a little too close to Saturn, close enough to reach the Roche limit and get destroyed by the immense tidal forces. The high ice content of the rings suggests that the moon may well have been massive enough to become either partially or completely differentiated, possibly even as large as Titan. The icy outer layers of the moon were stripped off as the moon got slowly ripped apart with the majority of the inner rocky material being engulfed by the mighty planet itself, while the icy debris from such a tidal encounter spread out in an orbit around the planet. It is very likely that the rings would have been more massive soon after their formation but that much of the ring material would slowly have coalesced under the force of gravity to form some of the moons we see today.

The rings are not only made of ice but also of gas. Data from the Cassini probe shows that the ring system has its own atmosphere composed of molecular oxygen. The gas comes from ultraviolet energy from the Sun striking the water ice particles and ejecting oxygen and hydrogen molecules from them. In another process, energetic ions from the moon Enceladus also strike the ice particles and release different molecules of oxygen and hydrogen. The result of both processes is a very rarefied atmosphere that surrounds the ring system.

Earlier in our journey around the Solar System we looked at flying through the asteroid belt. We took the risk and went for it. But then the distribution of material in the asteroid belt was really quite rarefied. What about the rings of Saturn? Would it be possible to fly through them safely? If you had to fly through them then the best route to take would be one of the many gaps like the Cassini Division or the Encke Gap. This was the approach taken by the Cassini spacecraft before entering into orbit around Saturn; it was directed through the gap between the F and G rings. The particles in the ring system vary from pieces smaller than a grain of sand to some that are as large as a house, so if a spacecraft were to try and fly through one of the rings there would be a pretty good chance of being hit. If one of the smaller particles were to impact then in all likelihood some damage would be sustained, though it shouldn't be catastrophic; a collision with one of the larger pieces would spell an end to the journey, however. It would take a brave and perhaps foolhardy adventurer to ignore the gaps and try to fly through one of the rings.

With all these particles of varying size in orbit around Saturn you might think the ring system will slowly disappear over time, but we have already seen how the shepherd moons keep the system intact. The Cassini Division is a 4,800-kilometre-wide region between the A and B rings that seems devoid of ring particles, although images from Voyager revealed particles within the gap but with a much lower density. The inner edge of the Division is maintained by Mimas, one of Saturn's many shepherd moons. Ring particles at the inner edge of the Division which are part of the B ring experience a 2:1 resonance with Mimas, so for every one orbit of Mimas, the ring particles orbit twice. This means that the pull of gravity from Mimas on the ring particles slowly builds, making their orbits unstable and resulting in a sharp cut-off of ring particles. In the rest of the Division are many thinner ringlets of less dense material, and among these are more gaps, such as the Huygens gap on the inner edge of the Division.

The Encke Gap can be found within the A ring and measures just 325 kilometres wide. This gap is the result of the small moon known as Pan which is on average 28 kilometres in diameter and which orbits in the gap. Any ring particles that drift into the ring in front of Pan will have their velocity slowed as the moon's gravity tugs on them, causing them to fall into the inner edge of the gap, whereas those that drift into the gap behind Pan will be accelerated, causing them to be ejected out of the gap towards the outer edge.

There are many other gaps and divisions within the rings. Some have explanations, others do not. But for the most part they are kept in place by gravitational effects from Saturn's many moons.

There are many wonderful structural features that can be seen in the rings that are also the result of the moons. Studies of the A ring revealed strange propeller-shaped disturbances along its outer edge, and it was soon discovered that tiny moonlets no more than 100 metres in diameter were orbiting within the ring debris and creating the disturbances. The fainter and more distant F ring was discovered by Pioneer 11 and is another great example where tiny moons have a big impact on the appearance of the ring. The ring is only about 300 kilometres wide in places and appears as two rings, with one spiralling around the other. It is kept in check by the two shepherd moons Prometheus (on the inside edge) and Pandora (on the outer edge), and it is their gravitational influence that seems to have shaped the beautifully dynamic nature of the ring. As Prometheus orbits Saturn, its most distant point takes it into the ring, causing it to steal some of the ring material and in the process induce disturbances in the ring that appear as knots and kinks. Pandora too seems gravitationally to disturb the ring particles, which are thought to be no larger than smoke particles, so that between the two moons and other possible unseen moonlets within the ring they have taken on a dynamic nature that has not been seen in a ring system anywhere else in the Solar System.

Not all ring features are driven by moons, however. Around the periphery of the B ring, for example, are vertical structures, some extending as high as 2.5 kilometres above the ring's main plane. Images from Cassini detected shadows cast by the structures when the angle from the sunlight was low, in just the same way that long shadows are cast on Earth when the Sun is low in the sky. There is no moon that seems to be creating these features so for now their origin remains a mystery.

Another feature whose origin seems to be eluding scientists is the surreal spokes that were first discovered by the Voyager probes in 1980. Until then it was assumed the ring system was a purely gravitational phenomenon, but this discovery of spokes which radiated out from the planet like the spokes on a bike wheel could not be explained. They appeared as dark patches when the Sun was behind the camera yet bright when illuminated from behind. What really perplexed scientists was the way they seemed to stretch across a number of rings yet retain their broad shape as they moved around the rings. Orbital mechanics should make this impossible: dust or ice particles that orbit nearer to Saturn would move faster than those further away so that any shape formed by them would disappear as soon as it had formed.

The particles that make up the spokes are now thought to be microscopic ice particles no more than a micron (one millionth of a metre) or so in diameter that have been elevated above the ring plane and suspended there by electrostatic repulsion. This happens because the particles in the spokes and the ring particles are thought to have the same polarity, which produces a repulsive effect. Because the particles are electrically charged they are also subject to the conditions and movements within Saturn's magnetic field, so as the field rotates with the planet, the particles in the spokes get moved too. The spokes seem to be much more common when sunlight is striking the rings from edge-on, so around the time equivalent to Saturn's spring and autumn equinoxes. During the summer and winter months they are almost absent, which suggests some link to seasonal variation. But for now the mystery of the ghostly spokes remains.

The Saturnian magnetic field that seems to drive the spokes is about twenty times weaker than the magnetic field on Jupiter but operates in the same way. Like Jupiter, Saturn has a liquid hydrogen core and it is currents within this core that seem to generate the field. As with all planets with a magnetic field, it deflects the solar wind from the Sun towards the poles where it accelerates particles already existing within the field causing them to produce beautiful aurora displays. They have been seen around Saturn's north and south polar regions.

Other features can be seen in the polar regions, such as the hexagonal cloud pattern at the north pole which is driven in part by a polar vortex which is itself the effect of solar heating and the movement of gas in the atmosphere. The atmosphere of Saturn is made up almost entirely of molecular hydrogen, a little over 96%, and just over 3% helium; the rest consists of various different elements including ammonia, methane and ethane, which are the main components of the clouds in the upper atmosphere. Visually, the atmosphere of Saturn has an appearance not too dissimilar from Jupiter's, although the banding and other features are much more subtle. The clouds responsible for the detailing in the upper atmosphere are mainly made of ammonia crystals, while the clouds lower down are thought to be made up of water and a chemical compound of ammonia, hydrogen and sulphur known as ammonium hydrosulfide. Ultraviolet radiation from the Sun is partly responsible for the creation of some of these chemical compounds, which are transferred around the planet through circulation cells like the high and low systems found in the weather on Earth.

The bands of clouds on Saturn were only discovered in the latter stages of the twentieth century with the advent of space exploration. Earth-based telescopes were simply not up to the job of detecting them. But now, with advances in optics, image processing and manipulation, the features and structures can be observed from Earth. The subtle system of bands is occasionally punctuated by storms that appear as white ovals against the pale cream background. A major storm seems to erupt every thirty years or so and, because their appearance resembles the Great Red Spot on Jupiter, they have been given the rather imaginative title of Great White Spots. They only seem to occur in the northern hemisphere of the planet and generally appear around the time of the summer solstice. It is quite likely that a similar event takes place in the southern hemisphere summer solstice (which is the winter solstice in the northern hemisphere), but owing to the orientation of the rings at this point, observation from Earth is challenging at best. The correspondence to the summer solstice suggests that the storms are related in some way to the amount of sunlight incident on the planet. The incoming solar radiation warms gas lower in the atmosphere, making it expand and causing its density to decrease, and because its density is lower than the surrounding gas, it rises. The white spots that we see are the huge atmospheric upwellings that start as moderately sized spots before increasing in size, particularly along the longitudinal axis. The last ones that were seen, in 1990 and 2010, stretched out so far that they encircled the entire planet.

Strong winds on Saturn are responsible for driving the storms around the planet, and they're also responsible for the generation of the belts. When Voyager visited Saturn in 1980 it detected wind speeds in excess of 1,800 kilometres per hour, which is over 1,000 kilometres per hour faster than winds on Jupiter. We have already looked at one of the effects of the strong winds on Saturn, the hexagonal cloud pattern at the north pole. This strange shape, which was discovered by Voyager in 1981, measures just under 14,000 kilometres along each side making the whole structure larger than the Earth. The real driving force behind the cloud pattern is not the wind speed itself but a sharp change in wind speed with latitude. As Saturn revolves on its axis, the gases in the atmosphere around the poles move at significantly different speeds. This not only produces a turbulent flow, but vortices seem to form on the southern side of these boundaries. The vortices, which are spinning columns of gas, interact with each other and evenly space themselves out around the pole, leading to the generation of the hexagon shape. There seems to be a very specific set of requirements in terms of wind speed and other atmospheric conditions before polar hexagons will form, which is why there is no corresponding one at the south pole. Studies in laboratories have attempted to recreate the process by rotating a fluid in a container, with the outer portions rotating slower than the central region. Similar results are seen in these experiments: shapes form around the axis of rotation, though they have not always just been six-sided. Sometimes they appeared with four sides, five, seven, or even eight.

Instead of a hexagonal cloud feature, the south pole is marked by a very well-developed hurricane about 8,000 kilometres across which is much like those seen on Earth – apart from its size, of course. Saturn's south polar storm is the only storm (other than those on Earth) that has a significant eye that is encircled by rings of huge towering clouds estimated to be 50 kilometres taller than the surrounding clouds. These so-called 'eye-wall clouds' are formed in terrestrial hurricanes when moist air travels in towards the centre of the storm, usually as it forms over an ocean, before rising, condensing into clouds and then releasing significant quantities of rain around a central column of descending air which forms the eye of the storm.

What makes the Saturnian hurricane unique is that it does not move or drift around. It seems somehow to have anchored itself to the polar region. Nor has it formed above an ocean, and on Earth, the moisture from bodies of water like that are a key ingredient for their formation. Wind speeds around the periphery of the eye have been measured at around 550 kilometres per hour, but within the eye, things are pretty calm. The eye appears dark because it is devoid of high-level clouds, which means that by studying the eye of the storm in detail we can peer deep down into Saturn's atmosphere. Using thermal imaging techniques that record infrared radiation it is possible to see clouds which are much lower in the atmosphere, giving us an unprecedented view of the conditions there.

The conditions in the atmosphere that surround the features at the north and south poles preclude the formation of water ice crystals. With pressures at around 50,000 pascals (100,000 pascals is roughly equivalent to atmospheric pressure at sea level on Earth) and temperatures between minus 170 and minus 110 degrees, the conditions are right for the formation of ammonia ice crystals. These crystals make up the majority of the clouds at this altitude, much like high-level cirrus clouds on Earth, although these are made of water ice crystals; but as altitude decreases, the composition of the clouds changes. A descent through the atmosphere will result in an increase in atmospheric pressure, and once the pressure reaches around a quarter of a million pascals and the temperature reaches minus 50 degrees, water ice crystals start to form. At the top of this region there will be a mixture of water ice and ammonium hydrosulfide which turns to a more pure water ice composition lower down. Clouds formed even lower in the atmosphere are composed of water and ammonia in their liquid form – more analogous to the fluffy cumulus or rain-bearing nimbostratus clouds on Earth.

Further down in the atmosphere it becomes clear that Saturn is a gas giant made almost entirely of gas with no solid surface to land upon. As we saw earlier, the gas is primarily hydrogen with quantities of helium and other elements like ammonia. Eventually a liquid metallic helium layer would be reached, just like Jupiter, then a liquid metallic hydrogen layer, then finally the planet's core. From studies of the way Saturn interacts with its moons, ring particles and passing spacecraft, it has been possible to deduce that the core is around twenty times more massive than the Earth with a diameter of about 25,000 kilometres.

Temperatures in the core are thought to be a little under 12,000 degrees, but the planet radiates out into space about two and a half times more energy than it receives from the Sun. The extra energy is thought to come from a process known as the Kelvin-Helmholtz mechanism. The mechanism is not unique to Saturn and is one that is thought to drive heat generation on gas giant planets and on brown dwarf stars. These are stars whose mass is insufficient for nuclear fusion to have started in their core. Instead, as the Kelvin-Helmholtz mechanism explains, the surface of a brown dwarf star or planet slowly cools after formation resulting in a gradual decrease in pressure, causing it to shrink. This compression then causes the core to heat up, generating extra energy. But in Saturn's case the Kelvin-Helmholtz mechanism is insufficient to explain its energy output so there must be some other form of heat generation. One possibility is droplets of helium rain that fall in the lower levels of the atmosphere, and as they fall through denser layers they generate heat through friction. This theory also explains the lower than expected levels of helium in Saturn's upper atmosphere.

The gravitational kick from Saturn means there is no need to fly through its rings in order to continue your journey deeper into the Solar System. Instead, a glancing fly-by will alter your trajectory enough to send you on to the next planet of the mission, Uranus. But before you leave the Saturnian system behind there is one more thing to take a look at. One image taken by Cassini in 2013 captured a stunning solar eclipse, with Saturn blocking out the Sun. The light from the Sun illuminating Saturn from behind makes it look beautiful, while the silhouette of Saturn's disc is surrounded by a rim of light being refracted through its upper atmosphere. The rings look absolutely breathtaking, taking on a whole new appearance in the shot. While the inner rings look fabulous, it is the faint outer E ring highlighted by light coming from the Sun that frames the entire planet like a halo. But what really makes it a special photograph is one tiny, rather insignificant speck of light piercing through just on the inside edge of the E ring. That speck of light is home, planet Earth. So, before you leave, you should certainly try to catch a glimpse of that sight for yourself.

EIGHT

Icy Outposts

NEARLY 1.4 BILLION KILOMETRES on from Saturn you encounter the first of two more giant planets, Uranus. When I was at school I remember learning that the Solar System was made up of four inner rocky planets, four outer gas giants, and then Pluto. It was only later when my interest in astronomy grew that I discovered the seventh and eighth planets from the Sun are actually ice giants, not gas giants. This does not mean that the planets are simply big balls of ice, but that they differ in terms of their composition. As we saw in the last couple of chapters, Jupiter and Saturn consist almost entirely of hydrogen and helium. In fact well over 90% of their composition is hydrogen. But in the case of both Uranus and Neptune, hydrogen accounts for only about 20% of their makeup.

The other 80% comprises heavier volatile elements such as methane and ammonia which have low boiling points.

So the planets we see today are not made of ice, as their collective name suggests, but when they formed over 4 billion years ago they would more than likely have been icy. The chemicals are now in a state described as a 'supercritical fluid'. In this state, the pressure and temperature of them are well above the levels necessary for phase boundaries to exist, so for example a gas can also behave like a liquid and a liquid can also behave like a solid. Other planets exhibit these conditions too, as we saw with Jupiter earlier. The pressure and temperature at the surface of Venus, too, is well above the critical point of carbon dioxide and nitrogen so the atmosphere at the surface is said to be a supercritical fluid, just like the gas in Uranus and Neptune's atmospheres.

The discovery of these two planets was a major triumph for science, and it all began on 13 March 1781. In that year, Sir William Herschel was engaged in a series of observations to measure the parallax (apparent shift of position due to the movement of Earth) of fixed stars. While he was observing stars in the constellation Taurus he found an object that he noted as 'either a nebulous star or perhaps a comet'. He looked for it again on 17 March only to find that it had moved, so he concluded it must be a comet. Following his announcement of a comet discovery, many astronomers across Europe took time to study the new object, but the absence of a nebulous coma or tail soon made it clear that the comet was in fact a planet. Herschel acknowledged this: 'by the observation of the most eminent astronomers in Europe, it appears that the new star, which I had the honour of pointing out to them in March 1781, is a primary planet in our Solar System'.

Over the following years the movement of Uranus around its

orbit was carefully observed and it was found that it seemed to be wandering from its expected path. These perturbations could only be caused by the gravitational influence of another, more distant planet. By applying Newtonian mechanics to the orbit of Uranus, the French mathematician Urbain Le Verrier was able to calculate the possible location of the undiscovered planet, giving astronomers an area of the sky to focus on. The discovery of Neptune in 1846 has since been attributed to both Le Verrier and John Couch Adams, a British mathematician and astronomer who had been working independently of Le Verrier.

For hundreds of years we knew about Mercury, Venus, Mars, Jupiter and Saturn as they were bright and easy to see from Earth. For an equal length of time astrologers have used the movements of the planets to predict the future of people living on Earth. These horoscopes, which are still sadly commonplace, did not of course take account of the impact of Uranus and Neptune – until, that is, they were discovered and all of a sudden they were affecting people's lives too. Your journey thus far around the Solar System has taught you many things, and one of them is bound to be that human beings are pretty insignificant when it comes to matters of the Universe. Why should all these incredibly beautiful and amazing planets affect us personally? There is no evidence that planets can influence our lives in the way astrologers would have us believe. So next time you read a newspaper and see your horoscope you would do well to avoid it, or at the very least take whatever advice it contains with a big pinch of salt. Certainly at this moment you'd be better off focusing your attention on Uranus as you approach it – it's what you've come all this way for after all.

With an equatorial diameter of 51,120 kilometres, Uranus is the third largest planet in the Solar System, after Jupiter and Saturn.

Its average distance from the Sun is 2.8 billion kilometres and it takes a little over eighty-four years to complete one full orbit. What makes this orbit interesting is the manner in which Uranus moves around the Sun. Most planets, the Earth included, orbit the Sun in an almost upright fashion. To quantify that statement, all the planets orbit the Sun roughly in the same plane. You can visualize that by imagining the Sun at the centre of a giant sheet of paper and the planets moving around along that plane. The axis of rotation of most of the planets is almost upright, although as we saw in an earlier chapter, for Earth it is tilted by 23.5 degrees when measured against the plane of the Solar System, or the sheet of paper in the analogy. Most of the planets are near enough upright in the way they rotate, but Uranus is tilted over on to its side so that its axis of rotation is 97.7 degrees. In a sense, it is rolling around the Solar System.

This arrangement gives Uranus a rather strange set of seasons, unlike any other planet in the Solar System. During the times of its equinoxes, when the length of day equals the length of night, the Sun is overhead at the equator giving the familiar experience of day and night, but the solstices – on Earth, the longest and shortest days of the year – are very different. At one solstice, one of the poles points towards the Sun, giving that hemisphere almost forty-two years of constant sunlight – although at this distance the light from the Sun is 400 times weaker than it is on Earth. Following a gradual move through the equinox the other pole is presented to the Sun which then enjoys forty-two years' daylight while the opposing pole is plunged into darkness for the same duration. Interestingly, however, although the poles receive a greater amount of sunlight, the equator has the higher average temperature. The reason for this is not known, but it must have something to do with the redistribution of heat within the atmosphere.

When Uranus is experiencing the summer solstice in either of its hemispheres the movement of the Sun would be quite alien to any visitor to the surface of any of its five major moons (remember, there is no solid surface on Uranus). All these moons share the same axial tilt as Uranus, which is one of the reasons why it is believed they formed out of a rotating accretion disc of material around the young planet. It is the rotation of objects that makes the Sun follow a path across the sky during the period of a day, but from any of the major moons of Uranus, the Sun would just follow a circular path around the celestial pole, never rising or setting, just gradually moving across the sky as the eighty-four-year-long orbit progresses.

We define the tilt of Uranus with reference to the plane of the Solar System, and using the definition from the International Astronomical Union (IAU) of the north pole as the pole which is 'above' the plane of the Solar System. Uranus, like Venus, rotates in the opposite direction to the rest of the planets and is said to have retrograde rotation. Uranus, however, rotates prograde. But quite why it is 'rolling' around the Solar System is not known. One theory looks to a period of time during the formation of the planets when Jupiter and Saturn experienced an orbital resonance where Saturn completed two orbits for every one orbit of Jupiter. This configuration had a small but cumulative effect on Uranus which may have been sufficient to affect its formation. Perhaps the most popular theory is a more simplistic explanation: it, like Venus, was struck by an Earth-sized object early in the history of the Solar System, knocking it over on to its side. Even the subtle ring system of Uranus and its family of moons orbit the planet around the equator so they spend half their time above the plane of the Solar System and half their time below it, adding some credence to the impact theory. Unfortunately there is little evidence on Uranus of such a significant impact as there would be on a rocky body. It may be that probing the planet's interior will help us to understand the mystery, but for now that is what it remains.

As you've just left behind the stunning ring system of Saturn, the system of rings around Uranus will seem really quite disappointing. They were first suggested by William Herschel in 1789 but his observations seem to lack support: no one else referred to them again for nearly 200 years. That said, he did describe the Epsilon Ring very well, including its colour, appearance and relative angle to the planet. The first confirmed detection came in March 1977, the rings' presence inferred when observations were made of the occultation of the star SAO 158687. Occultations are events that occur when one object blocks another from view, so in effect a solar eclipse is an occultation where the Sun is blocked from view by the Moon. In the case of the observation of Uranus, the star was predicted to be blocked from view by the planet, but what surprised astronomers is that it faded from view five times before and five times after it passed behind the disc of Uranus. The conclusion was inescapable: Uranus was surrounded by a system of rings that were too faint to be seen directly. When Voyager 2 visited the Uranian system in 1986 it sent back the first direct images of the rings, showing that there were eleven in total. Then data from the Hubble Space Telescope recorded in 2005 showed two more outer rings, bringing the number to thirteen.

The rings of Uranus, which span around 60,000 kilometres of space, differ quite substantially from those around Saturn in that the ring particles are small and dark, usually just a few fractions of a millimetre to up to a metre across. There is a wonderful measure in astronomy known as the bond albedo which describes the total amount of radiation that arrives at an astronomical object which is returned to space through the scattering of light (whereas 'albedo', mentioned earlier, refers only to sunlight). The ring particles of Uranus have a bond albedo of just 2%, which suggests they probably consist of ice rather than rock, although their dark nature means they cannot be pure water ice. They are more likely to be a mixture of ice and a dark as yet unidentified material, perhaps even organic in nature. By studying the way the appearance of the rings changes with different angles of light it is possible to determine more about the nature of these ring particles.

During certain lighting orientations the Epsilon Ring, the one Herschel claimed to have observed, appears quite red in colour, suggesting that there may be amounts of dust within the system. Along with being the brightest of the rings, the Epsilon Ring is also among the thinnest with an estimated thickness of just 150 metres. There are nine inner main rings and two dusty rings a little further away from the planet, including the Gamma Ring, which when observed in back scattered light becomes brighter than the Epsilon Ring. Beyond those two dusty rings are the two outer rings that were discovered by Hubble which differ quite significantly from the inner system in many ways, not least because at 17,000 kilometres and 3,800 kilometres across they are quite wide. When the rings were studied in the near infrared with the large multi-mirror Keck telescope on top of Mauna Kea, the outer ring known as the Mu Ring was not detected but the Nu Ring was picked up. This suggests the Mu Ring is bluer in colour and so more than likely composed almost entirely of microscopic dust, unlike the Nu Ring, which is expected to contain more ice.

Unlike the rings found elsewhere in the Solar System, the rings around Uranus are thought to be quite young, having formed after the planet. The most popular theory on their formation is that the particles were once part of a moon that was shattered by high-velocity impacts during the earlier history of the Solar System and the resulting debris formed the rings we see today. This is supported by observation of the ring particles which are in orbits that are comparable with the orbits of most of the moons, and, just like the moons of Saturn, they act to keep the ring system in shape.

There are twenty-seven moons in orbit around Uranus and they are broken down into three groups: thirteen inner moons, nine irregular moons and five major moons. Titania and Oberon were the first of the major moons to be discovered, by William Herschel in 1787. Over sixty years passed before the next two were spotted, in 1851. They were named Ariel and Umbriel and were found by William Lassell, a beer brewer and astronomer, using one of his home-made reflecting telescopes. It was almost another century after that, in 1948, when Gerard Kuiper discovered the last of the major moons, Miranda. Voyager 2's fly-by in 1986 revealed ten more moons, and a further moon was identified later, after closer inspection of the Voyager photographs. The remaining moons were all discovered through the Hubble Space Telescope and Earth-based telescopes in the decades that followed. Unusually, all the moons are named not after Greek or Roman gods but after characters from the plays of William Shakespeare and the works of the poet Alexander Pope.

Miranda is the closest of the major moons, and inside its orbit are the thirteen inner moons. Puck and Mab are the outermost of that group, and the latter is the source of the particles making up the Mu Ring. All the inner moons are closely related in some way to the rings of Uranus, be it providing material for the rings like Mab or acting as shepherd moons to one of the rings, just as Cordelia and Ophelia do for the Epsilon Ring. Although the inner moons are small – only two are over 100 kilometres in diameter – they constantly perturb each other in their orbits, causing them to shift over time. As a result of this, collisions between the moons are not uncommon, which is one of the mechanisms that seeds the rings with fresh material.

Miranda, Ariel, Umbriel, Titania and Oberon, whose sizes range from 470 kilometres to around 1,500 kilometres, dominate the Uranian system of moons. It is unclear exactly when they formed but it is likely they either came out of the accretion disc that surrounded Uranus soon after it formed or they are the result of the impact that is thought to have knocked the planet on to its side. All the moons except Miranda are made up of almost equal amounts of rock and carbon dioxide ice mixed with ammonia. Miranda is composed of significant quantities of water ice with amounts of darker silicate rock and shows a surface with evidence of intense geological activity. It is covered in canyons and grooved features that criss-cross the moon which formed as the crust stretched during tectonic movement. Most of this geological activity has been driven by internal tidal heating caused by orbital resonances with other moons. For example, in the early history of the Solar System, Miranda was in a 3:1 orbital resonance with Umbriel, where it completed three orbits for every one of Umbriel's. After a few million years Miranda lost the resonance with Umbriel but its orbit changed significantly, causing it to become highly eccentric. With an eccentric orbit the forces exerted on Miranda from Uranus varied significantly over time, so it was constantly being stretched and squeezed in different directions, leading to the internal tidal heating.

The subject of internal heating on Uranus is interesting but for very different reasons. The sheer mass of the giant planets tends to be a contributing factor to the amount of heat they generate internally, so that they radiate more heat than they receive from the Sun. In the case of Uranus, the amount of heat it produces is significantly less than the heat it receives from the Sun. Temperatures recorded in its atmosphere are as low as minus 224 degrees, making it the coldest of all the planets. Neptune is similar in size and composition to Uranus yet its heat output is over two and a half times greater.

The reason for the low temperatures on Uranus still eludes us. One possible theory looks to the impact that dislodged Uranus from its upright orientation and in doing so may have liberated much of the heat from the core. Alternatively it may simply be that there is no vertical transfer of heat through the atmosphere. The transfer of heat around a planetary atmosphere is driven by convection, which relies on the variation of density within the material either as a result of compositional differences or based on thermal properties. These variations or gradients can slowly disappear over time which leads to a reduction in the ability to transfer heat around the planet, unless gradients exist in other regions to continue the transfer. In the case of Uranus, it is quite probable that heat transfer is being limited by a process known as double diffuse convection, where different regions of the atmosphere have different density gradients and hence different convective properties, which may actually inhibit the transfer of heat away from the core.

We can tell a lot about the interior structure of Uranus by studying how it moves around the Sun, how the moons move around it, and how it interacts with passing spacecraft. From this information we can infer a lot about the mass and the distribution of matter, and knowing that it has an equatorial diameter of just over 51,000 kilometres we can deduce that its density is 1.27 grams per cubic centimetre. This makes it slightly more dense than Saturn, which has a density of 0.68 grams per cubic centimetre. This information tells us that almost 80% of the total mass of Uranus is made of ice.

Underneath the hazy blue disc is a world separated into three distinct regions: the atmosphere, a mantle of ice and a rocky silicate core. The core is estimated to be just over 10,200 kilometres across, making it a little smaller than the Earth, but the conditions are very unEarthlike: the temperature is of the order of 5,000 degrees and the pressure equivalent to the surface pressure at Earth multiplied by a factor of 8 million. Under these pressures, the material in the mantle which is described as ice is actually a rather exotic form of hot, dense liquid of ammonia and other volatile elements that in many ways act as an ice.

The atmosphere is rich in methane molecules, but with the immense pressures at these low levels the methane molecules are ripped apart into carbon and hydrogen atoms. The carbon atoms crystallize under the extreme conditions into what scientists at the University of California once described as 'diamond rain'; there may even be an ocean of liquid diamond underneath the mantle of the planet. If you were to travel from that liquid diamond ocean upward then the liquid would slowly turn into gas as you reached higher levels in the atmosphere. Like all of the giant planets, Uranus has no solid surface, so instead, when discussing altitudes in the atmosphere, a datum is taken at the point where the atmospheric pressure equals the pressure at the surface of the Earth – this is how we calculate the nominal diameter of gas planets, which as we know for Uranus is 51,120 kilometres. Above this level the atmosphere extends out by several thousands of kilometres.

The general composition of the atmosphere is mostly molecular hydrogen and helium and quantities of methane which is responsible for the blue/green colouration of the planet. Unlike Jupiter and Saturn, cloud features in the atmosphere of Uranus are few and far between; indeed when the Voyager 2 probe arrived it detected only ten cloud systems over the entire planet. One of the key driving forces in the production of cloud is heat, which causes parcels of gas to move around the atmosphere. The low levels of heat received from the Sun and the low level of internal heating mean that Uranus is cold so clouds are sparse. When observed visually, the planet lacks any spectacular features, even from this close up as you coast past.

When Voyager 2 visited, the southern hemisphere was experiencing summer so was presented towards the Sun; owing to its trajectory, it did not get to study the northern hemisphere. It found the southern hemisphere to have a few large-scale features, such as a bright cap at the south pole and subtle dark bands that circle the equator. Between the polar cap and the dark equatorial band is a bright belt centred at a latitude of minus 47 degrees. It's the brightest feature on the planet and has been dubbed 'the collar'. The collar and cap are now believed to be dense regions of methane clouds which are around 30 kilometres below the zero-kilometre altitude level.

Your arrival in the system coincides with the equinox so you can see a different season and how that has affected the features. Not too surprisingly, the southern collar seems to be disappearing and there is a hint of one forming around the northern pole at a latitude of around 50 degrees, so there is evidence that the south polar region is darkening while the northern pole is brightening. This supports the theory that the features are clouds which form as the hemispheres of the planet slowly warm and cool with the small amounts of solar radiation that arrive. Other clouds seem to be forming in the northern hemisphere too as it turns to face the Sun, but there appear to be differences between them and their southern hemisphere counterparts. Clouds in the southern hemisphere seem to be larger and persist for longer while those in the northern hemisphere seem to be smaller and brighter, perhaps because they lie at a higher altitude and are reflecting more of the incoming radiation.

Studying the clouds in the atmosphere that appear at various different altitudes allows us to determine the wind speeds and directions, and profile how they change with altitude. The wind speeds at the poles are zero but increase to nearly 250 metres per second (equivalent to around 900 kilometres per hour) at a latitude of around 60 degrees where they blow in a prograde direction, the opposite direction as the rotation of the planet. Moving further towards the equator, the winds decrease in speed again to a point at around 20 degrees latitude where once again they are absent because it is at this latitude that the temperatures are at their lowest. The wind direction around the equator is reversed, or retrograde, and blows in the same direction to the rotation of the planet with speeds in excess of 100 metres per second.

The wind speeds on Uranus seem high by Earthly standards but they are nothing compared to those found on the final major planet on the *Kaldi*'s journey, Neptune. Getting from Uranus to Neptune may in your mind seem like something of a hop, but in reality it will take you another three years to traverse the distance between these two outermost planets.

At its closest approach to Earth, Neptune is 4.3 billion kilometres away. It is incredibly difficult to visualize the true vastness of the Solar System, but one way to do it is to scale the distances down to more manageable numbers. If we say that 1 million kilometres in the Solar System equals 2.8 kilometres on Earth, and place the Sun in central London, then on this scale the nearest planet from the Sun, Mercury, would be about 163 kilometres away in Birmingham and Venus would be in Plymouth. Earth, which is usually about 150 million kilometres from the Sun, would be residing on the Isle of Man and the red planet Mars would be in Prague, no doubt having a highly cultural experience. The distances between the planets then start to become quite extreme. Jupiter is 2,200 kilometres away in Egypt, and the ringed planet Saturn is in Iraq. Uranus, the planet you have just visited, would be over in the Far East, in China, and Neptune, the final planet on your journey, would be on the other side of the world, in New Zealand.

Space travel is not for the impatient.

For the entire duration of this leg of the journey you will be able to see Neptune clearly up ahead as a bright blue disc against the velvety blackness of space. Its blue hue is subtly different to the more aquamarine appearance of Uranus, both, as we have seen, the result of the chemical composition of the planets.

Neptune orbits the Sun at an average distance of 4.5 billion kilometres and takes 164 years to complete one orbit. Despite the fact that it is an ice giant like Uranus and about four times the diameter of Earth there are similarities with our home planet. Neptune rotates once on its axis every sixteen hours and six minutes so, because this is only about eight hours shorter than Earth's own rotation, a day on Neptune is similar to a day on Earth. The axis of rotation is similar too: Neptune's axis is tilted with respect to the plane of the ecliptic by 28 degrees while the axis of Earth is just over 23 degrees. This means that Neptune has seasons much like those on Earth, although owing to the much longer orbital period, the seasons last for forty-one years.

As is the case with Uranus, the formation of Neptune is still a subject of scientific debate. One problem with the standard theory, which posits that the planets condensed out of the accretion disc in their current orbits, is the presence of minor bodies in the vicinity of Neptune: the formation of a major planet like Neptune in this location would surely have also swept up these objects. It is also believed that at these distances the disc out of which the planets formed would not have had a sufficient density of matter. Instead it is likely that the two ice giants formed closer to the Sun where the matter density was higher and then migrated out to their existing positions after the accretion disc cleared – a theory known as the Nice (pronounced 'Nees') Model.

Neptune does resemble Uranus, but you will notice some prominent features as you draw closer, particularly the Great Dark Spot, which is reminiscent of the Great Red Spot on Jupiter. It too is a raging anticyclonic storm but somewhat smaller, measuring just 13,000 kilometres by 6,600 kilometres – though that's still big enough for one Earth to fit across its widest axis. The storm was discovered back in 1989 by Voyager 2 when it visited the Neptunian system; but when the Hubble Space Telescope turned its gaze on the planet in 1994, the storm had vanished. Instead, the HST spotted another storm but this time in the planet's northern hemisphere, so the term 'Great Dark Spot' is now more generally used to describe the dark spots seen in the atmosphere of Neptune rather than any specific one. It is around the first spot, known as GDS-89, that the fastest wind speeds in the Solar System have been recorded, at a staggering 2,400 kilometres per hour – a little higher than the maximum cruising speed of Concorde.

The spots seem to lack any features or clouds, suggesting they may be vortex-like structures and that what we are seeing is the top of a hole through the methane clouds that allows us to peer deeper into the atmosphere, perhaps even as far down as the troposphere. White clouds of methane ice crystals similar to Earth's cirrus clouds seem to form around the periphery of the spots. The methane clouds appear to last longer than the spots themselves, and this persistence means we can pinpoint where previous spots might have existed before the holes either closed up or became obscured. The HST observation showing that GDS-89 had disappeared suggests that in contrast to the Great Red Spot on Jupiter, which has been visible for over 400 years, the spots on Neptune are short-lived, lasting for just a few months or years. It may be that they simply dissipate as they near the equator, or perhaps some other as yet unknown process leads to their demise.

The clouds and wind speeds seen on Neptune are due in part to internal heating. Neptune receives less than half the amount of energy from the Sun that Uranus receives and it is 1.6 billion kilometres further away, yet their temperatures are comparable. At the zero altitude level measured where the atmospheric pressure is 1 bar, the temperature is minus 201 degrees compared to minus 197 degrees at the same level on Uranus. The only explanation for this is that Neptune must be producing and distributing heat internally, with current estimates of 2.6 times the amount of energy received from the Sun. The origin of this internal heat is unknown, particularly as Uranus is so similar to Neptune in many ways yet produces significantly less heat. The most popular theory is that the heat is simply left over from the formation of the planet, but as we have seen this does seem to be contradictory to the observations of Uranus.

All weather systems are driven by heat, be it external (from the Sun) or internal. The source of heat from inside Neptune is generally considered to be constant year on year, although there will be an almost imperceptibly slow decrease as the planet cools. The amount of heat certain areas receive from the Sun, however, will alter with the changing presentation of the planet to the Sun. One of the ways we perceive this is in the changing seasons. Evidence of seasonal changes and large-scale movements of atmospheric gases have been detected on Neptune with concentrations of methane and ethane that are up to 100 times higher around the equator, suggesting there is a general upward movement of gas around the equator and subsidence of gas around the poles. The tropospheric temperatures at the south pole also hint at seasonal changes, with temperatures a few degrees warmer there than elsewhere on the planet. During these southern hemisphere summers the warmer temperatures are sufficient to turn frozen methane into a gas which escapes into space. This causes the southern pole to appear a little brighter than the surrounding regions, but as the Neptunian year progresses and the north pole starts to present itself to the Sun, then it will brighten as it starts to warm. We saw an identical process on Uranus which resulted in darker and lighter polar regions dependent on the seasons.

The similarities with Uranus don't end there. Neptune's atmosphere is made up almost entirely of hydrogen and helium with traces of methane which, owing to the way this greenhouse gas absorbs red light, gives the planet its striking blue colour. There are four regions to the Neptunian atmosphere: the troposphere, the stratosphere, the thermosphere and the exosphere. The clouds we see tend to occur at different altitudes within the troposphere and their altitude will determine the nature of the clouds. In the upper regions the pressures are sufficiently low for methane clouds to form, and as the pressure increases at lower levels clouds of hydrogen sulphide and ammonia are found. In the lower levels, the pressure increases to around five times the pressure at the surface of the Earth; the clouds here are made of ammonia sulphide and water. As the pressure increases further in the lowest reaches, there are dense clouds of hydrogen sulphide.

The atmosphere itself accounts for about 10% of the overall mass of the planet; the core and mantle comprise the other 90%. The amount of material in the mantle is equivalent to about fifteen times the material that makes up the Earth, but instead of being a rocky lump it is a hot, super-dense liquid which is made up of ammonia and water. The water is not in a state we are familiar with in our oceans though, as the conditions cause the hydrogen and oxygen to dissociate into hydrogen and oxygen ions, making the liquid highly conductive. At lower levels in the mantle the high pressure causes the methane to separate into hydrogen and carbon, leading to the creation of diamond hail – a process very similar to that found deep in the layers of Uranus. It is also hypothesized that even higher pressures further into the mantle lead to the formation of a diamond ocean, with diamond bergs floating around. There is little direct evidence to support this so for now it remains a wonderfully romantic theory, but if these entities do exist they could well be exactly like the diamonds we know and covet on Earth.

Deep under the mantle is the core of Neptune. It is a silicate rock with a mixture of iron and nickel and a total mass about 1.2 times that of the Earth, and of comparable size. At this depth, buried beneath thousands of kilometres of liquid and gas, the pressure reaches a crushing 7 million times that on the surface of the Earth, and temperatures exceed 5,000 degrees – almost as hot as the visible surface of the Sun.

Also surrounding Neptune is a ring system somewhat

reminiscent of the one around Uranus. There are just three main rings in this system: the inner Galle Ring, the Le Verrier Ring and the outermost Adams Ring. They span a 21,000-kilometre-wide region of space around Neptune and are composed of thousands of ice particles that are covered in carbon material, giving them a dull red colour. It was once thought, as a result of observations of stellar occultations, that the rings had gaps in them: the stars flickered in and out of view as the planet approached but not once Neptune had passed. When Voyager 2 visited the system it revealed that instead of there being gaps in the rings, they simply had a rather clumpy structure. The outermost ring, the Adams Ring, is a great example of this with five confirmed arcs of higher density. It is thought that this clumpy nature is a direct result of the gravitational influence of the tiny moons that orbit within the ring system. In the case of the arcs in the Adams Ring, the nearby moon Galatea is responsible. The constant tug and disruption from the moons seems to be causing changes in the rings over geologically short timescales, and it is likely that within a few centuries some of the ring features we can see today may well vanish.

In addition to Galatea there are thirteen other moons in orbit around Neptune, and they fall into two categories. Nearest to the planet are the seven regular moons which orbit in the same direction in which Neptune rotates on its axis, and all are within its equatorial plane. This is in contrast to the irregular moons which tend to orbit further away in a retrograde or backward fashion and with orbits inclined to the equatorial plane with a highly elliptical shape. There is one exception to this broad categorization of the moons, Triton. It should be a member of the outer irregular group because of the direction of its orbit but, unlike all the other irregular moons, it lies close to Neptune.

Triton was discovered just seventeen days after Neptune by British astronomer and brewer William Lassell. It is by far the largest of the Neptunian moons and constitutes 99% of the mass of the material in orbit around the planet, and with a diameter of 2,700 kilometres it is the seventh largest moon in the Solar System. This makes it large enough to have evolved with a broadly spherical shape. It is not unusual for moons to orbit in a retrograde direction around the planet – three other moons orbit Neptune in the same direction. There are other moons around Jupiter, Saturn and Uranus which have retrograde orbits, but they are all much more distant from their planets. Triton is unusual because it orbits at a distance of just 354,000 kilometres.

It is not possible for any moon that orbits a planet in a retrograde motion to have formed out of the same part of the nebula that the planet formed. Taking into account its composition and retrograde orbit, it is thought Triton may be a captured Kuiper Belt object. Since its possible capture from the belt, the orbit of Triton has become almost perfectly circular. One force that is often responsible for the circularization of an elliptical orbit is the tidal force, but in the case of Triton that is unlikely to have provided sufficient drag to affect the orbit that much. It is more likely that drag from the debris disc surrounding Neptune slowed Triton sufficiently to make its orbit circular. Tidal interaction still plays a part in the evolution of Triton's orbit as it constantly tugs on the moon, slowing it further. Over time, perhaps even within the next 4 billion years – about the same amount of time that has elapsed since the Solar System formed – Triton will get so close to Neptune that it is likely to be destroyed by tidal forces and form a new ring system.

It takes Triton 5.8 days to complete one orbit of Neptune but it also takes 5.8 days to complete one rotation on its axis, so in the same way that our Moon keeps one face pointing towards Earth, so Triton keeps one face pointing towards Neptune. This is known as synchronous rotation and is found in many moon–planet relationships. In fact, take a quick trip down to Triton's surface in the science-busting RSU, because from this vantage point things are going to appear quite strange. From the surface of Earth we are used to seeing things rise in the east and set in the west, but because of the synchronous rotation, Neptune is hanging motionless in the sky above you. It looks big too, doesn't it? It spans an area of sky about 8 degrees across, which is sixteen times the apparent size of the full Moon back on Earth. The axis about which Triton rotates is currently at a 40-degree angle to the plane of the orbit of Neptune, so as the pair of them orbit the Sun, the poles of Triton point alternately at the Sun. The changing orientation of the moon to the Sun means that it experiences seasonal changes, and each of those seasons last for about forty years.

Perhaps the most interesting aspects of Triton are its atmosphere and some of the features you can see on the surface. That surface is composed of frozen water, nitrogen and carbon dioxide; measurements of its density suggest the entire body is 45% ice with the remaining 55% comprising rock, giving it a composition somewhat similar to Pluto. The high quantity of ice on Triton's surface means that it is highly reflective. In fact it reflects 80% more light than our own Moon. This is one of the contributory factors in its discovery: a lower reflectivity would mean it was a lot less visible. On the surface there are many different types of feature visible, from ridges and troughs to icy plains and plateaus. What is noticeable by their absence are craters, which seem to be a common feature in the Solar System. This indicates the surface of Triton is geologically young – anything from 5 million to 50 million years old.

One of the more obvious things you'll spot are the dark streaks that were first discovered by Voyager. They were quickly identified as being connected to geyser-like eruptions of nitrogen gas; the streaks were sub-surface dust caught up in the eruptions. The geysers tend to be more active at points on the surface where the Sun lies directly overhead, suggesting that solar heating, however weak, has something to do with their origin. One theory suggests that solar radiation penetrates the thin icy crust, warming the rocky surface below. Pressure below the ice builds until it reaches a critical level, leading to the geyser-like eruptions that can reach heights of up to 8 kilometres. Many of the eruptions are thought to last for anything up to one Earth year, throwing up sufficient dust to create streaks that stretch downwind for anything up to 200 kilometres. The icy nature of these events has led to them being dubbed cryovolcanoes, although they differ from other cryovolcanoes which are driven by internal heat rather than heat from the Sun.

The discovery of the streaks also tells us that Triton has an atmosphere, because if there is wind, there must be an atmosphere. It is not dense like the atmospheres on some of the moons in the outer Solar System – its tenuous nature is one of the reasons the cryovolcanic eruptions can reach altitudes of 8 kilometres. We can tell a lot about the atmosphere of moons and indeed planets by observing the way light is extinguished as stars pass behind them. The light from any object that passes behind our own Moon, for example, extinguishes instantly rather than fades and flickers out of view. This tells us that there is no appreciable atmosphere surrounding the Moon. Observations of a star that passed behind Triton showed that its atmosphere was denser than that recorded by Voyager some eight years earlier, and the surface temperatures too were a little warmer, by no more than 5%. These important observations told us at the time that its average temperature was on the increase, perhaps as it heads towards a warmer summer.

Nitrogen is the main component in the atmosphere, with small amounts of methane and carbon monoxide. The majority of the nitrogen has come from the gradual sublimation of surface ice. From the surface level, the temperature slowly decreases from about minus 237 degrees with increasing altitude to the tropopause at a height of around 8 kilometres. Above our own tropopause, and indeed on many other objects in the Solar System, is a stratosphere where temperatures start to increase with height rather than decrease and gases are separated into strata dependent on their temperature. Triton has no stratosphere; instead, the tropopause changes to a thermosphere whose characteristics are like the stratosphere's, with an increase in temperature with altitude as a result of ionization from the incoming ultraviolet radiation. It differs from the stratosphere due to the way gases are separated into different strata based not on their temperature but on their molecular mass (determined by adding the mass of each of the atoms and multiplying by the number of atoms in the molecule). The thermosphere reaches altitudes of around 950 kilometres, and above that is the exosphere. An exosphere is the outermost layer of any atmosphere, marking the boundary with space. Its molecules are still gravitationally bound to the planet but do not tend to interact with each other in the way most gases do.

Triton is by far the most interesting of the moons, which is why you paid it a short visit, but there are thirteen others in orbit around the planet which deserve a little attention. In the same category as Triton are Nereid, Halimede, Sao, Laomedeia, Neso and Psamathe. Nereid orbits Neptune in a prograde direction but its orbit is highly elliptical, taking it from 1.4 million kilometres to the furthest point in its orbit at 9.7 million kilometres. Its eccentricity is high. We say that 0 is a perfectly circular orbit and 1 is known as a parabolic escape orbit, which in effect means that an object with an eccentricity of 1 would escape the system. Nereid's eccentricity has a value of 0.75, which leads to the conclusion that it was either an object captured from the Kuiper Belt, like Triton, or was previously a member of the inner regular moons. If it was the latter then it is likely that the appearance and subsequent gravitational interaction of Triton could very easily have disrupted its orbit, turning it into the one we see today.

The regular moons are closer to Neptune than the irregular moons, and in order of distance from the planet they are Naiad, Thalassa, Despina, Galatea, Larissa, S/2004 N1 and Proteus. The smallest of them all is S/2004 N1, which is around 15 kilometres in diameter. Contrary to the suggestion in its name, it was discovered in 2013; the '2004' comes from the fact that images from the Hubble Space Telescope taken in 2004 showed the moon, so that was the year in which it was first recorded. All of these moons are in some way related to the ring system. We have already seen that Galatea is responsible for the features in the Adams Ring; Naiad and Thalassa orbit between the Galle and Le Verrier Rings, while Despina is a shepherd moon for the Le Verrier Ring only. All of the inner moons are likely to have formed after a chaotic and disruptive period caused by the capture of Triton. Its immense mass would have sent these inner moons into turmoil, ejecting some and more than likely causing the orbits of the rest to vary wildly, leading to collisions and their eventual destruction. Over time, as Triton settled into its present orbit, the system calmed down and the debris that was left closer to Neptune would have slowly coalesced under the helpful force of gravity, thus forming the small inner system of moons and rings you can now see.

There are a couple of other objects in the Neptunian system which are best described as companions rather than satellites. They are Neptune's trojans and can be found at the L4 Lagrangian point (we looked at these earlier and saw how they are places where gravitational forces in a three-body system balance). The L4 point lies 60 degrees ahead of Neptune and it is here that six trojans are found, with a further three at the L5 point, 60 degrees behind Neptune. All of them must by definition orbit the Sun over the same period of time as Neptune and follow broadly the same orbit. The discovery of the third trojan, known as 2005 TN53, was significant because its orbit was found to be tilted with respect to Neptune's at an angle of about 25 degrees. This tells us that there is a high likelihood of a greater number of trojans at this point, almost like a swarm of flies.

Your path through the Neptunian system does not take you near these trojans. Unfortunately, their dim red appearance makes them almost impossible to pick up against the blackness of space without any form of optical aid. The route has taken you close by Triton, however, the largest moon of Neptune, skimming over the top of the ring system, and on a close fly-by of Neptune itself. Neptune then gives you one final boost of acceleration, taking your speed to a little over 17 kilometres per second. At that speed you could fly around the Earth in just thirty-eight minutes.

Not only does your final planetary fly-by increase the *Kaldi*'s speed, it also adjusts its trajectory and allows you to set the course for the ship's ultimate destination. Not Pluto, not Proxima Centauri (the nearest star system to our own), but to a star in the constellation of Libra known as Gliese 581 at a distance of twenty light years. It of course takes twenty years for light to reach you from that system, and at the *Kaldi's* current speed it will take about 352,000 years to get there. However, it is on this leg of the journey that the ion engine will be beneficial. Firing it over a long period of time will produce a gradual increase in velocity and maybe get the ship up to around 25 kilometres per second, which will reduce the journey time to a mere 239,000 years. Obviously the vast distances involved would make the journey something of a suicide mission for any human inhabitants on board the ship, so fortunately once you reach interstellar space the RSU will allow you to return home once more to the comforting familiarity of planet Earth, leaving the *Kaldi* to complete the mission unmanned. There are alternative methods that could vastly reduce the journey time, such as nuclear pulse propulsion, where a series of nuclear explosions propel a spacecraft forward at a maximum speed of about 1,000 kilometres per second. This would cut the journey time to Gliese 581 to just six years, but at the moment it is only an idea.

Gliese 581 is a fairly normal red star about a third of the mass of the Sun, but it has been chosen because it's surrounded by a family of three planets, one of which, Gliese 581c, is believed to be 21 million kilometres from the star itself. It created excitement when it was discovered in 2010 because it orbits the star in the habitable zone, which is the region within which conditions could be just right for liquid water to exist on the surface, and maybe even right for life to evolve.

But before that, and before you become the first human being to enter interstellar space, we must explore the outermost regions of the Solar System.

NINE

Into the Abyss

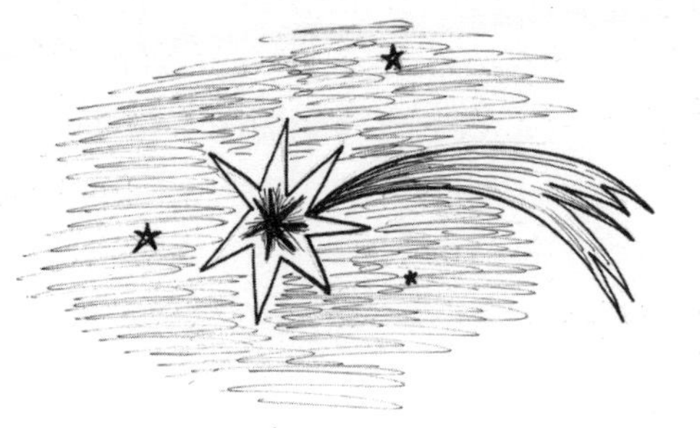

IT HAS NOW BEEN seventeen years since you left Earth, and in that time you have witnessed first-hand the awesome power of our nearest star, the Sun, and the effect it has on the unwary space traveller. You have journeyed to the inner planets, gazed upon the cratered landscape of Mercury and experienced the hostile conditions on Venus. You have walked on the surface of Mars, a world that has been visited by a number of unmanned space probes, and become the first human to leave footprints there. You then had plenty of time for reflection as Jupiter slowly grew bigger in the *Kaldi*'s viewing window. Seeing both it and Saturn close up was truly magnificent; the latter's rings at close range were spellbinding. An intimate cruise past the mysterious icy worlds of the giants Uranus and Neptune was a fitting way to round off the most incredible trip, a true once-in-a-lifetime experience. But the journey has not quite ended yet.

Behind you lie the familiar planets of the Solar System, but what lies ahead? The discovery of Uranus and Neptune was a wonderful triumph for mathematics and for science in general. You will recall that Uranus was discovered by chance when William Herschel was studying stars in the constellation Taurus, and that Neptune was finally detected following mathematical analysis of the orbit of Uranus. Careful study of the orbits of Uranus and Neptune showed further perturbations. It seemed there might well be another planet out there in the depths of the Solar System gently tugging at these two icy giants.

It was left to astronomer Percival Lowell, founder of the Lowell Observatory in Flagstaff, Arizona, to try to find it. He set about the search and soon homed in on a number of possible locations for the unidentified ninth planet, which was soon dubbed 'Planet X'. Sadly he died without knowing his team had captured images of it on two separate occasions. Clyde Tombaugh was also working at the Lowell Observatory and between 1929 and 1930 had been comparing photographic plates of the same part of the sky that were taken a few nights apart. After a year of painstaking work, Tombaugh spotted a faint moving object, which was eventually called Pluto. Its discovery was thought to complete the picture of the Solar System.

The case for Planet X was reopened in 1978 with the discovery of Charon, the faint moon of Pluto, which allowed astronomers to calculate accurately the mass of Pluto. When it was found to be just 0.2% of the mass of the Earth it was immediately clear that Pluto was far too small to cause the observed perturbations in the orbits of Uranus and Neptune. Then the Voyager fly-by of Neptune in 1989 allowed that planet's mass to be calculated. The Voyager data reduced the giant's mass by just 0.5%, and reapplying this new value to the orbit of Uranus nicely accounted for the perturbations. This finally laid to rest the need for a Planet X: Uranus and Neptune were moving exactly as they should be.

Finding Pluto, it seems, was just a stroke of luck. Diligent scientific observation shows that Planet X never existed and was proposed simply to account for unexplained variations in the orbit of Uranus. But does this mean that our Solar System is complete and we will never discover another major planet out in the depths? The chances of such a discovery seem very slim. The Voyager and Pioneer probes are now heading out into interstellar space and neither has detected the presence of a previously unidentified planet.

That said, in recent years there has been a slow revival of the Planet X concept due to a number of observed characteristics in the outer Solar System. The most recognized of these is the so-called Kuiper Cliff. The Kuiper Belt is found beyond the orbit of Neptune and can be thought of as a larger version of the asteroid belt between Mars and Jupiter, although its composition includes many more frozen elements such as water, ammonia and methane, owing largely to the greater distance from the Sun. Pluto and a number of the other 'trans-Neptunian' objects are now considered to be Kuiper Belt objects. It is reasonable to assume that the existence of Kuiper Belt objects would gently decrease with distance from the Sun, but in reality the belt seems suddenly to terminate at a distance of about 48 astronomical units (some 7 billion kilometres – as you may recall, 1 astronomical unit, or AU, is the average distance between the Earth and Sun). Some scientists believe that this 'Kuiper Cliff' is the result of an unidentified planet orbiting the Sun at this distance, perhaps the size of Earth. Such an object would not only explain the abrupt termination of the belt, it would also account for why a few objects seem to have been ejected into different orbits.

Far beyond the Kuiper Belt is another region of icy objects known as the Inner Oort Cloud, thought to be a part of the theorized Oort Cloud out of which comets originate. In 2003 the first suspected member of the Inner Oort Cloud was discovered and named Sedna. It's an icy object that has a diameter of about 1,000 kilometres, but what makes it remarkable and of great interest to astronomers is that its highly elliptical orbit takes it from its closest point to the Sun, at just 11.3 billion kilometres, to an incredible 140 billion kilometres, which is 940 times further than the Earth–Sun distance. When you compare these figures to the average distance from the Sun to Pluto, which is 5.9 billion kilometres, then you realize just how far away Sedna is. At that distance it is only accessible for observation from Earth for a mere fraction of its 11,400-year-long orbit. Outside that short window, it cannot be seen. What has really excited the scientific community is the nature of its orbit and what may have caused it. Perhaps the passage of a nearby star dislodged it from its orbit within the Inner Oort Cloud; maybe it was exiled to the outer regions of the Solar System by other stars that formed with the Sun; or has its orbit been changed by the presence of another planet-sized object that is currently beyond the range of our detection?

The discovery of another suspected Inner Oort Cloud object, 2012 VP113, has brought astronomers a step closer in terms of understanding these objects with highly elliptical orbits. 2012 VP113 has been found to come as close as 11.9 billion kilometres to the Sun at its perihelion and as far away as 67 billion kilometres at aphelion. If a few more of this type of object can be found then close examination of their orbits will allow us to deduce more about their natures and where they have come from.

Unlike Sedna and 2012 VP113, Pluto is thought to be a Kuiper Belt object, and since 2006 has been classed as a minor planet because it is not gravitationally dominant in its orbit. It is joined on its 247-year orbit of the Sun by five natural satellites, the largest of which is Charon with a diameter of 1,200 kilometres, compared to the 2,368-kilometre diameter of Pluto. Charon is less massive than Pluto at 1.5 sextillion kilograms (Pluto's mass is an estimated 13 sextillion kilograms – 0.2% of the Earth's mass), but because their masses are of the same order of magnitude, the point at which their gravitational force balances is above the surface of Pluto. Where objects are an identical mass then this point, which is known as the barycentre, would sit exactly halfway between the two objects, but if one is more massive then the point moves towards the higher-mass object. For all the major planets in the Solar System, the mass of the planet is so high in comparison to the mass of the moon that the barycentre is, if not close to the centre of the planet, then at least somewhere within the body of the planet.

It is very difficult to learn more about Pluto from space exploration because of its mass. Sending a spacecraft so far out in the Solar System means it has to travel fast to oppose the gravitational pull of the Sun, but because Pluto has such a low mass, any spacecraft that does venture out that far cannot be captured by its gravity so instead continues straight past it. Missions like New Horizons, which launched in 2006, will only get a couple of days of close-up observation at best – a brief opportunity to study the surface detail. Thankfully, even at these huge distances the Hubble Space Telescope has been able to record enough details to give us a moderate understanding of the minor planet's nature.

Before the HST, some of the first attempts to map Pluto's surface involved very careful study during moments when Charon eclipsed Pluto. As the moon passed in front of the planet it slowly blocked out brighter and darker regions which caused a change in the overall brightness of the system. By studying this change in brightness, a very rough map could be produced of the surface features. The result was a surface that changed quite significantly, not just in brightness – or rather reflectivity – but also in colour, ranging from a dark grey through to a dark orange and even white. Over the first few years it was observed, the northern polar region brightened and the southern region darkened, suggesting seasonal changes as a result of its axial tilt. Its colour too seems to have changed, becoming a little more red than before, possibly as a result of chemicals from the surface sublimating into the atmosphere. Spectral studies of the surface reveal it is made up almost entirely of nitrogen ice with small portions of methane and carbon monoxide.

By studying the way Charon moves in its orbit around Pluto and knowing the minor planet's volume, we can determine its overall density, which is around 2,000 kilograms per cubic metre. This suggests that it is composed of around 70% rock and 30% ice. Due to heat released from the radioactive decay of elements, the ice would melt, allowing them to separate from the rock and giving Pluto a differentiated structure. If it has evolved with a structure like this then the core will be around 1,700 kilometres in diameter and be surrounded by an ice mantle over 300 kilometres thick. Some scientists have suggested that any current radioactive decay may have caused the complete melting of the ice in the core–mantle border, which could produce a liquid sub-surface ocean, but further study is needed to understand if this could actually exist.

The surface is composed almost entirely of nitrogen ice with small amounts of carbon monoxide and methane. The observed seasonal changes have a relationship with the thin atmosphere of Pluto. While they are affected by the axial tilt of the planet, they are also affected by the distance between Pluto and the Sun. At its perihelion, Pluto is 29.6 times further from the Sun than the Earth; at its aphelion it is 48.8 times further away. When Pluto is nearer the Sun, more of the surface ice sublimates and thickens the atmosphere; when it is at its most distant and the temperatures are at their lowest, the gas in the atmosphere turns straight back into a solid in a process known as deposition. The transfer of material between the surface and the atmosphere is one of the main reasons why the surface appearance of the planet changes over time.

Much of what we know about the atmosphere of Pluto has come from the observation of stars as they disappear behind the planet. If there was no atmosphere then stars that pass behind Pluto would simply vanish in the blink of an eye; if there was an atmosphere the light from the star would gradually fade. The average atmospheric pressure at the surface of Pluto varies from between six hundred thousandths of the pressure experienced at the surface of the Earth to two hundred and forty thousandths, with the greater atmospheric pressure experienced when Pluto is nearer the Sun.

Arriving at Pluto fills you with a mix of emotions. The demoted planet, now only of minor status, will always hold a special place among the countless smaller chunks of rocks in the Solar System. With the exception of the New Horizons mission, Pluto has really not been explored much, so this is a great opportunity to take a look around this icy outpost.

As you take your first tentative steps on the surface you catch a glimpse of the Sun in the sky. Surprisingly from this distance the Sun is still quite bright, obviously much fainter than it is from Earth but still uncomfortable to look at – about 200 times brighter than the full Moon appears on Earth. You have arrived at Pluto while it is nearing its most distant point from the Sun and, because the ellipticity of its orbit is so great, its changing distance has an impact on its surface temperature. This ranges between minus 240 degrees and minus 218 degrees, which when compared with the coldest temperatures recorded on Earth of minus 92 degrees seems rather chilly. At this temperature many of the gases in the atmosphere will freeze on to the surface like frost. The scene that greets you, therefore, is like a cold, hard-frost morning back at home, the surface a glistening white.

Moving around is, as you might expect, not too dissimilar to moving around on the Moon. The easiest and most efficient method of manoeuvring in such low gravity (about a twelfth that of the Earth) is by hopping along. Unless you are feeling particularly brave, though, do not put too much effort into your leaps. Most people jump with an initial velocity of 4 metres per second, so with an escape velocity of around 1.2 kilometres per second there is no danger of you floating off into space, but you would certainly get to a decent height. If you can jump about a metre on Earth then a leap with the same effort on Pluto would get you soaring to a height of about 30 metres, so you'd need a head for heights.

And you need to be careful not to lose your footing on landing. The terrain is uneven with loose material, and there are craters dotted around; add that to the frostiness of the landscape and it's easy enough to slip over. But if you do slip over you won't fall as instantly as you do on Earth, it'll be a more graceful event. You'll have plenty of time to put your hand out to break your fall – which is essential if you want to ensure you do not damage your space suit on any of the jagged rocks that pepper the surface.

Of all the places you have visited and walked on, Pluto appears to be the most alien. The eerie light, the frostiness of the surface and your superhuman ability to jump over houses have made this one final excursion to remember. As you prepare to return to the ship you look back towards the Sun and pause, realizing that Earth is somewhere out there in the inky blackness. In a moment of homesickness a solitary tear wells in your eye, but the gravity is so weak here that it does not roll down your face.

There are a number of theories to explain Pluto's origins. One of the earlier ideas suggested it used to be a moon of Neptune and that the arrival of Triton in the Neptunian system dislodged Pluto from its orbit. That now seems unlikely because, although Pluto gets closer to the Sun than Neptune, at no point do their orbits cross. The full story started to reveal itself in the early 1990s with the discovery of more small icy and rocky bodies beyond the orbit of Neptune. These trans-Neptunian objects seemed to share many properties with Pluto, not least in terms of approximate size, composition and even orbital properties, and Pluto is currently accepted as the largest member of the group.

We have already looked briefly at the Kuiper Belt and the Kuiper Cliff, which seems to mark a sudden decline in Kuiper Belt objects at a distance of about 48 astronomical units from the Sun. There are currently well over 1,000 known Kuiper Belt objects, most of which are made of ice and rock, and it is thought that these are tiny bodies left over from the formation of the Solar System that never quite formed into planets. As the outer Solar System was forming, a huge influence was exerted on the minor bodies by the large gas giants as they were settling into stable orbits, and it is this which is thought to have influenced the formation and current structure of the Kuiper Belt. Jupiter and Saturn eventually settled into orbits with a 5:2 resonance so the two mighty planets meet three times during every five orbits of Jupiter. The consequences of this planetary resonance were felt throughout the outer Solar System. They disturbed Uranus and Neptune. The orbit of Neptune in particular became a little more eccentric, sending it out further into interplanetary space. This had a big impact on the planetesimals that are now part of the Kuiper Belt, sending them out further into space and altering their orbits into much more eccentric ones, like that of Pluto. It is possible too that a great number were ejected off into space, perhaps even reducing the population to well under half of its original total.

In much the same way that Jupiter's immense gravity dominated the evolution of the asteroid belt, then, Neptune has affected the development of the Kuiper Belt beyond it. The presence of Neptune in its current orbit seems to be maintaining certain features, including perhaps the inner and outer boundary of the main belt (and it extends from the orbit of Neptune for another 3.7 billion kilometres). Objects that orbit along the inner boundary tend to have an orbital resonance with Neptune of 2:3, so that for every three orbits of Neptune, the Kuiper Belt objects (KBOs) complete two orbits. Pluto is one such KBO which orbits the Sun twice for every three orbits of Neptune, and in recognition of this, any other object with the same resonance with Neptune is called a Plutino. Another resonance with Neptune may be keeping the outer boundary formed, the so-called Kuiper Cliff, so that KBOs at this distance complete two orbits for every one of Neptune.

Between the two boundaries, in the region known as the Classical

Kuiper Belt, the gravitational effects of Neptune are not felt and the orbits of the KBOs are left largely undisturbed. Within the Classical Kuiper Belt there are two distinct groups of objects. The first is known as the 'cold population': the name does not reflect their temperature, it comes from the fact that their movement is somewhat analogous to the movement of molecules in a cloud of cool gas. They have nearly circular orbits which are constrained broadly to the plane of the ecliptic and have a different composition, making them appear more red than the other group. The 'hot population' have very different orbits which are much more elliptical and inclined to the ecliptic by as much as 35 degrees. The origin of the two groups is thought to be different, too: it is believed the cold population formed in their current position whereas the hot population may have formed closer to the Sun, perhaps in the vicinity of Jupiter, but were forced further out as the giant planets settled into their current orbits.

Beyond lies the Kuiper Cliff, which as we saw earlier is defined by a 2:1 resonance with Neptune. Its boundary is understood; what is not understood is why there seems to be few objects beyond it. We saw one possible explanation earlier: a large, currently unseen planet may be gravitationally restricting other objects. But it may simply be that there was insufficient material for the belt to form into a planet.

At your current speed, and with Pluto sitting at a point in its orbit where it is at the inner edge of the Classical Kuiper Belt, it takes us another eighteen months to transit the belt and reach the Kuiper Cliff – not that this would be noticeable to you. In fact you may not even have spotted a KBO in all the time you were in the belt: they would have been very dimly lit and almost undetectable. The void that seems to exist beyond the Kuiper Cliff is largely unknown. The two Voyager craft were directed out of the Solar System above and below its plane, as was Pioneer 11; Pioneer 10 was the only craft that was sent out along the plane. There is not a lot of direct evidence for objects at this distance from Earth simply because objects are not illuminated much by the Sun and are therefore difficult to detect. And if there were any more decent-sized planets out here, any craft or space traveller would be very lucky to be in the right place at the right time to see them.

From here on, the ion engine is going to be fired up to increase velocity and speed your journey through the outer reaches of the Solar System. The speed slowly increases, but it still takes twenty-three years to get to the edge of the Solar System, where interplanetary space ends and interstellar space begins, soon after which you'll depart the *Kaldi* for the very last time. This edge is defined by the point where the influence of the Sun is matched by the influence of other stars, and understanding exactly where this is means understanding the nature of the interstellar medium – the matter that exists between the stars in our Galaxy. This matter is made up of a mixture of gas (mostly hydrogen and helium), dust and radiation, and it is within this medium that a bubble exists which surrounds the Sun. The bubble is the heliosphere, which is generated from the pressure exerted by the solar wind, and the point where the pressure from the solar wind is balanced by the 'wind' from other nearby stars marks the outer limit of the Solar System. On 25 August 2012, Voyager 1 became the first man-made object to pass through this point and enter interstellar space, at a distance from the Sun of 18.1 billion kilometres.

The exact shape and structure of the heliosphere is still not fully understood, and with the passage of the Voyager spacecraft many more questions were raised than were answered. The solar wind which generates the heliosphere leaves the Sun at speeds of up to 750 kilometres per second, and it is the interaction of the solar wind and the interstellar wind that drives the structure. The wind from the Sun is composed of a magnetic field and electrically charged particles known as ions, but because the Sun rotates on its axis it induces a spiral-shaped ripple through the Solar System. When you visited the Sun at the start of your journey we looked at the eleven-year solar cycle, but something else happens every eleven years too: the magnetic field of the Sun reverses, so that with a change in polarity the north pole becomes the south and vice versa. This produces disturbances in the spiral-shaped heliospheric current sheet which leave Earth susceptible to cosmic ray strikes. As the Earth orbits the Sun, it is usually protected from cosmic rays by the current sheet; it's only vulnerable when it is moving through from one wave to another. During field reversal, Earth is likely to be much more at risk from cosmic ray strikes. This is a particular concern for space travellers and spacecraft outside the protection of the Earth's atmosphere – although the *Kaldi* has some protection, of course, from the superconducting magnets producing your very own magnetic field.

As the Sun itself travels through the interstellar medium at a speed of about 83,700 kilometres per hour, it and the region of space around it slams into the interstellar medium causing the solar wind to slow to subsonic speeds, generating a shockwave. This point is known as the termination shock, and at this point the solar wind gets compressed like a group of people trying to walk through a dense crowd. The density of material in the interstellar medium is actually pretty low, with just 10 million molecules per cubic centimetre. But even with that low density it takes around

13.3 billion kilometres for the strength of the solar wind to drop sufficiently to be slowed to speeds lower than the speed of sound. This reduction in speed produces the shockwave. The distance of the termination shock from the Sun is not consistent: Voyager 1 measured its distance (from determination of the speed of the solar wind and its temperature) to be 94 astronomical units, while Voyager 2 measured it to be 84 astronomical units. This apparent discrepancy is a result of the motion of the Sun through the space.

Beyond the termination shock, in the heliosphere, the solar wind is slowed even more and the increased interaction with the interstellar medium causes turbulence. The turbulent nature of the heliosheath (the outer region of the heliosphere) means that the speed of the solar wind varies – indeed Voyager 1 detected regions where the solar wind unexpectedly dropped to zero, although it later increased again. The heliosheath is thought to be comet-shaped, extending to about 100 astronomical units from the Sun in the direction of its movement but stretching out many times further in the downwind direction. It is contained by the heliopause, the region that marks the edge of the Solar System, and it is here that the solar wind, which left the Sun travelling at hundreds of kilometres per second, is finally brought to a halt. Until August 2012 this was a purely theoretical boundary but its existence, as we saw in chapter 3, was confirmed by Voyager 1. Its presence was marked by an increase in cosmic rays which are usually blocked by the heliopause, a reduction in temperature, and a change in direction of the magnetic field.

Just like the shockwave that formed at the termination shock inside the heliosphere, one theory has suggested it may be possible for a shockwave to form in front of the heliosphere as it moves through the interstellar medium. However, unlike the solar wind, which travels at supersonic speed, the relative motion of the interstellar medium as we move through it is subsonic, at just 83,700 kilometres per hour. This sounds supersonic by normal standards, and indeed the speed of sound at ground level on Earth is only 1,234 kilometres per hour, but the speed of sound varies with temperature and density. It is much higher in the more rarefied interstellar medium and our Solar System is simply moving too slowly to form a bow shock. According to NASA's Interstellar Boundary Explorer there is more likely to be a bow wave than a bow shock – more like the structures formed in front of a boat slowly moving through the water.

From this point on, you are in interstellar space. You have finally passed through the heliopause at a distance from the Sun of about 18.1 billion kilometres and it has taken you a little over forty-two years to get here. According to your flight plan the next port of call is, or more accurately *might be*, the Oort Cloud. This theoretical cloud surrounds the Solar System between 5,000 and 55,000 astronomical units from the Sun. Even with the slow but continual acceleration from the ion propulsion system it will take 1,500 years to get there, so the *Kaldi* will have to carry on without you. The time has finally come to return home.

Fortunately, thanks to the science-busting nature of the RSU, the journey home is nothing more than a flick of a switch. Once safely back on Earth the journey is not yet over, and before you can finally be reunited with your family and friends you must spend time in quarantine. Having journeyed around the Solar System and visited strange new worlds for the first time, great care must be taken to ensure that you have not brought back anything that could pose a risk to life on Earth. This same caution has been exercised ever since the astronauts returned from the very first Moon landing. Unlike them, however, your adjustment to life back on Earth is much more straightforward, as you have no issue reacclimatizing to the influence of our planet's gravitational field because you have been experiencing simulated gravity for the majority of your journey. After a series of tests and examinations you can finally leave the quarantine facility for an emotional reunion with your loved ones. It's been a long, fascinating journey but, as you take in your first breaths of fresh air, it feels great to finally be home.

The *Kaldi* continues on without you to the Oort Cloud, which is thought to be a massive halo of icy bodies that surrounds the entire Solar System, its outer members so distant that they are well on their way to being a quarter of the way towards the nearest star, Proxima Centauri. It is thought to have two distinct regions: a doughnut-shaped inner cloud known as the Hills Cloud and a spherical outer cloud which is only loosely gravitationally bound by the Sun. Within the two clouds there are believed to be several trillion icy planetesimals, each of them measuring about a kilometre in size. The only evidence for the existence of the cloud comes from the observation of comets, icy visitors to the inner Solar System that occasionally grace the skies of Earth.

Studies have shown that there are two groups of comets: the short-period comets that orbit the Sun over relatively brief periods of time (less than 200 years) and the long-period comets whose orbital period is greater. The short-period comets tend to come from the Kuiper Belt but the long-period comets, whose orbits can be many thousands of years, are believed to originate in the Oort Cloud.

Other than the time it takes the comets to orbit the Sun, there is no fundamental difference between the two. Both are a mixture of rock and ice, although the long-period comets tend to have more ice than their short-period counterparts. At the centre of a comet is a nucleus that usually measures just a few tens of kilometres across and is often likened to a dirty snowball. You may be able to remember as a child scooping up a handful of snow only to find you had picked up a fair amount of soil and stone with it. Just like your snowball, the nucleus of a comet is primarily ice but contains varying quantities of rock too. While the comet remains in the outer reaches of the Solar System its ice stays solid, but when some kind of disturbance sends it towards the Sun the temperature increases and things start to change. Because the nucleus of the comet is so small with an almost negligible gravitational force associated with it, the nuclei do not retain an atmosphere, so with the low pressures that come with an environment like this, ice that gets heated turns straight into a gas rather than a liquid. With the sublimation of ice, the nucleus gets surrounded by a halo of dust and gas. These halos are known as the coma of a comet, and when the pressure from the solar wind pushes against them they can extend for millions of kilometres, forming the comet's trademark tail. It is a common misconception that the tail of a comet stretches out behind it as it whooshes through space. As we have just seen, the reality is that the tail always gets 'blown' downwind from the solar wind, which means that it always points away from the Sun.

Spectroscopic studies of comets and robotic space exploration have shown that many comets contain ammonia and other elements that are the building blocks for amino acids and proteins, which are key to the evolution of life. This important discovery reveals that the seeds that brought life to Earth may well have arrived by comet. When the Earth was young it would have suffered numerous bombardments from asteroids and cometary nuclei bringing with them a variety of life-supporting chemicals. When the theory was first suggested there was concern that the delicate compounds may well have been destroyed in such an impact. To counter any concern, experiments were conducted where high-velocity projectiles coated in organic compounds were fired at metal plates. The experiments simulated the forces and conditions experienced during a comet impact on Earth and found that the compounds survived. It is even just possible that the energy released during the impact may have become a catalyst for chemical changes to kick-start the evolution of life.

By studying the orbits of the long-period comets it has been possible to build a profile for the likely place they came from. The appearance of long-period low-inclination comets allows us to put an estimate on the inner doughnut-shaped Hills Cloud at between 2,000 and 20,000 astronomical units. Comets that appear with a very long-period orbit – such as Hale-Bopp, the so-called 'Great Comet of 1997', which had a highly inclined orbit of 2,537 years – suggest the outer cloud structure ranges from about 20,000 to 50,000 astronomical units. When these comets appear they can be big and spectacular because they spend much of their life in the deep freeze of the outer Solar System rather than making regular visits to the Sun like their short-period cousins who shed quantities of ice on their orbits around the Sun.

If the Oort Cloud does indeed exist then, like most things in the outer reaches of the Solar System, there is some uncertainty as to where it came from and how it formed. The most popular theory for a number of years explains that the cloud formed out of the remains of the protoplanetary disc that surrounded the young Sun. It is quite likely that the cloud could have formed nearer to the Sun and gravitational interaction with the giant planets ejected its members into highly elliptical and distant orbits. More recently, studies have suggested that the cloud may have been formed as a result of an exchange of material between the Sun and the stars it formed with, since most stars are believed to form in hot young stellar clusters; over time, they drift apart to lead either solitary lives with families of planets or as members of binary or multiple star systems. With the slow but gradual separation of the stars, this latest theory suggests that they swapped material which eventually led to the formation of the very distant cloud of objects that surrounds our Solar System. It is also likely that some of the outer members of the cloud could still be interacting with nearby stars as they pass through the Galaxy.

There are other influences experienced by the Oort Cloud, not least of which are gravitational forces from the Galaxy itself. Just like planets in orbit around the Sun or moons in orbit around a planet, they are all subject to gravitational tidal forces. At the distance of the Oort Cloud the gravitational influence of the Sun is substantially weakened, so much in fact that the gravitational force of the Galaxy has played a more significant role in the evolution of the cloud. During its early formation, many of the objects within it would have had highly elliptical orbits but the galactic tidal forces have led these orbits to become circularized into the spherical cloud. It is also possible that the tides could dislodge objects from within the cloud, sending them in towards the Sun. The Hills Cloud is closer to the Sun so the galactic influence there is less. There has not been sufficient time to make the orbits circular.

The *Kaldi* emerges from its long stay in the Oort Cloud and continues the remainder of its journey to Gliese 581, just over twenty light years away in the constellation Libra. The majority of the journey will be through interstellar space and will involve travelling through the interstellar medium. As we learned earlier, the density of molecules in the medium is low. In liquid water there are 10 sextillion (that's a 1 with twenty-two zeros) atoms per cubic centimetre, but in the medium there is on average one atom per cubic metre. If the ship's fuel runs out then the *Kaldi* will continue on at the same speed and in the same direction unless acted upon by another force, which might perhaps come from the gentle nudge of a nearby star.

On arrival at Gliese 581, the Sun will appear as a pretty insignificant fourth-magnitude star. All objects in the sky, including the planets, are described in terms of their brightness by their magnitude, with brighter objects having a negative number and the faintest objects visible to the naked eye assigned a value of 6. From Earth, the Sun has a magnitude of minus 26, the full Moon is minus 13, Venus is minus 5 at its brightest and the most distant visible object, a galaxy known as UDFj-39546284, has a magnitude of 29, which is 500 million times fainter than the human eye can detect.

From the surface of the Earth, Gliese 581 is not visible to the naked eye; it requires a pair of binoculars to be able to see it. It was chosen as your mission's ultimate destination because of the possibility of alien civilization.

The star itself is nothing special. It's a red star, cooler than our Sun but like our Sun is still fusing hydrogen to helium deep in its core. It is much smaller than the Sun, though, with about a third of its mass, which accounts for its lower temperature and redder colour. Until April 2007 it received very little attention, but that was until the discovery of Gliese 581c, the second planet found to be in orbit around the star. What particularly interested astronomers about this planet was that it was the first exoplanet to be found orbiting a star within the habitable zone. This is the zone within which any planet in orbit with a suitably dense atmosphere can sustain liquid water at the surface.

581c is a planet that has a mass about 5.6 times the mass of the Earth. Its distance from Gliese 581 is 11 million kilometres, compared to the Earth–Sun distance of 150 million kilometres, but the lower temperature of the star puts the planet right on the edge of the habitable zone. At that distance from the star, it takes only thirteen days for it to complete one orbit so its year is considerably shorter than ours, but unlike Earth it is tidally locked with Gliese 581. As we have seen before, this means that just one face of the planet stays facing the star with the other remaining permanently in the dark. This has quite significant implications for the likelihood of liquid water on the planet. Any water that does exist is very likely to evaporate in the high temperatures of the daytime side and then freeze on the night-time side through deposition. Given the tidal locking on the planet, the atmosphere could reach a state where the entire water content of the planet has frozen solid on the night-time side. One way that water can be detected on planets around other stars is to study the light from the star as it passes through the atmosphere of the planet. In this way the existence of water vapour in the atmosphere gives itself away through the light from the star that it absorbs. This extraction of the light from the star behind is seen as dark absorption lines in the spectrum of the star. Unfortunately in the case of Gliese 581c, the orientation of its orbit does not pass in front of the star when viewed from Earth.

There are three confirmed planets in the system. One of them, 581e, is just 1.7 times the mass of the Earth. Unfortunately, this one orbits the star at a distance of just 0.03 astronomical units, which equates to 4.5 million kilometres from the parent star – over 53 million kilometres closer than Mercury orbits the Sun. At that distance it completes one orbit of the star in just over three days and is bathed in searing heat and doses of radiation that make life on the planet very unlikely.

All the planets around Gliese 581 were discovered using the radial velocity method of extrasolar planet detection. The technique relies on one simple principle which we have looked at already, the movement of objects around the barycentre, the centre of gravity of the two objects in question. To recap, if two objects are of equal mass then they will orbit each other around a point directly between them. If one of the objects is more massive, then this point will lie closer to it, and if it is significantly more massive then the point will lie almost at the centre of the more massive object. Even in a case like this, the more massive object will still orbit around the point although its movement will be tiny and barely detectable. In the case of planets in orbit around a star, the presence of the planet in orbit will cause the star to wobble a tiny amount.

It is possible to detect the tiny wobble or movement of the star by studying its light through a spectroscope, which as we have seen splits the incoming light into its spectrum. As the star moves, the absorption lines in the spectrum will be moved first towards the red end of the spectrum and then towards the blue end of the spectrum in an effect known as the red and blue shift. This concept is something we have all experienced as the Doppler effect, such as when an emergency vehicle passes with its siren sounding. As the vehicle approaches, its motion causes the sound waves to get bunched up, causing the pitch of the siren to increase, but as it continues past, the pitch changes and starts to decrease as the sound waves are stretched out again. It is similar for the spectrum of the star: its movement causes the light to get bunched up as it moves towards you, making the absorption lines appear to shift towards blue; as it moves away again, the light waves get stretched out so the lines shift back towards the red end again. If the shift in position in absorption lines is carefully observed and measured then it is possible to calculate the movement of the star, and from that information the mass of the object that is tugging it out of position can be worked out. The evidence of the presence of three planets in orbit around Gliese 581 is hidden in the subtle movements of these absorption lines.

Other techniques are used to discover planets around distant stars. One of the more popular ones involves studying not the spectrum of light from the star but its brightness instead. There are a number of reasons why the apparent brightness of a star may vary, but occasionally it will be because a small amount of its light is being blocked by something such as another star or perhaps even a planet. The tiny variations in brightness as a result of a planetary transit are almost imperceptible but their signature is unmistakable. As the planet transits the star there is a tiny reduction in brightness of the overall system as the planet blocks a tiny amount of starlight from reaching us, and there will then be a secondary dip when the planet passes behind the star. This second dip is detectable because the planet usually reflects a small amount of starlight, thus increasing the brightness of the overall system, but when it passes behind the star that reflected light is not seen. If this were all plotted on a graph to show light against time then the curve would appear like a hilly landscape as the light level changes, with the valleys representing the time when the planet and star are in alignment. The existence of more than one planet around a star is revealed in repeating patterns in the light curve.

In the future, new technology may allow us to take direct images of the surfaces of these so-called exoplanets. Orbiting space telescopes that utilize a technique known as interferometry will be able to combine the light from telescopes separated by thousands of kilometres to offer views with unprecedented resolution, revealing fine surface detail on distant worlds. For now, though, all we can do is wonder what the conditions are like. Sadly it will be quite a few years before humans can even think about setting foot on a world around another star as there will have to be some significant improvements in rocket propulsion and supporting technologies for that to happen.

When we do finally make that journey, are we likely to find new civilizations thriving on other worlds? The chances seem quite high. After all, the Universe is a complex system of billions and billions of galaxies, and each one contains billions of stars. In fact it is often said that there are more stars in the Universe than there are grains of sand on Earth. Our current knowledge suggests that water is plentiful in our Solar System, planets around other stars are not a rarity, and even organic compounds seem to be abundant, so the chances for life to evolve seem high.

In case there is life on one of the planets around Gliese 581 and the *Kaldi* comes into contact with it, a plaque has been attached to the side of the ship to show where it has come from. A similar plaque was attached to the Pioneer spacecraft, which used a map of nearby pulsars (rapidly rotating dense stars) to identify the location of our Solar System. In addition to that there's a map on the *Kaldi* of the Solar System and a depiction of the spacecraft coming from the third planet around the star, which is of course our home, planet Earth.

Even if the ship does not encounter alien civilization, the mission has been a great success. From the comfort and safety of home you will be able to look back on a fascinating voyage across the expanse of our Solar System. You have been closer to the Sun than any other human being, visited the hostile worlds of the inner planets, successfully negotiated the asteroid belt, saw the beautiful cloud tops of Jupiter and even flew through the raging hurricane that is the Great Red Spot. You have seen the awesome rings of Saturn and the subtle hues in the atmospheres of Uranus and Neptune, then explored the dark depths of the outer Solar System. After that you became the first human to leave the Solar System and enter interstellar space, before leaving the *Kaldi* to continue the voyage without you.

The journey required you to spend many years away from home, but the long-term space mission has been quite an adventure. Wanting to explore the unknown is an innate part of human nature, and just like our ancestors who set out to map every inch of the Earth's surface, to cross seas, climb mountains and race to reach the poles, we have continued the quest to understand the natural world, peering into and continually wishing to explore the dark depths of space. Yours has been an amazing experience, and your trusty spacecraft will continue on through interstellar space for all eternity.

Index

Index

Index

Index

Index

Index

About the Author

For the past four series, **Mark Thompson** has been one of the presenters on the BBC 2 award-winning show *Stargazing Live*, and most recently the resident astronomer on ITV's *This Morning*. When not gracing our television screens he is most likely writing about the sky, looking at it or flying through it (as a qualified pilot, not a human cannonball). He writes for a number of websites including Discovery News's Space pages and the Space Exploration Network, along with a variety of other publications. He is also a regular on Radio 5 Live.

Born in Norfolk, Mark has had a fascination with all things in the sky ever since he was a small boy and is proud to be a fellow of the Royal Astronomical Society.